Student Note-Taking Guide to accompany

AF881621

Fifth Edition

Human Biology

Daniel D. Chiras
Colorado College

JONES AND BARTLETT PUBLISHERS
Sudbury, Massachusetts
BOSTON TORONTO LONDON SINGAPORE

World Headquarters
Jones and Bartlett Publishers
40 Tall Pine Drive
Sudbury, MA 01776
978-443-5000
info@jbpub.com
www.jbpub.com
Jones and Bartlett Publishers Canada
2406 Nikanna Road
Mississauga, ON L5C 2W6
CANADA

Jones and Bartlett Publishers International
Barb House, Barb Mews
London W6 7PA
UK

Copyright ©2005 by Jones and Bartlett Publishers, Inc.

ISBN: 0-7637-3605-8

All rights reserved. No part of the material protected by this copyright may be reproduced or utilized in any form, electronic or mechanical, including photocopying, recording, or by any information storage and retrieval system, without written permission from the copyright owner.

Printed in the United States of America
09 08 07 06 05 10 9 8 7 6 5 4 3 2 1

Contents

How This Book Can Help You Learn

All of us have different learning styles. Some of us are visual learners, some more auditory, some learn better by doing an activity. Some students prefer to learn new material using visual aids. Some learn material better when they hear it in a lecture; others learn it better by reading it. Cognitive research has shown that no matter what your learning style, you will learn more if you are actively engaged in the learning process.

The Student Note-Taking Guide will help you learn by providing a structure to your notes and letting you utilize all of the learning styles mentioned above. Students don't need to copy down every word their professor says or recopy their entire textbook. Do the assigned reading, listen in lecture, follow the key points your instructor is making, and write down meaningful notes. After reading and lectures, review your notes and pull out the most important points.

The Student Note-Taking Guide is your partner and guide in note-taking. Your Guide provides you with a visual guide that follows the chapter topics presented in your textbook. If your instructor is using the PowerPoint slides that accompany the text, this guide will save you from having to write down everything that is on the slides. There is space provided for you to jot down the terms and concepts that you feel are most important to each lecture. By working with your Guide, you are seeing, hearing, writing, and, later, reading and reviewing. The more times you are exposed to the material, the better you will learn and understand it. Using different methods of exposure significantly increases your comprehension.

Your Guide is the perfect place to write down questions that you want to ask your professor later, interesting ideas that you want to discuss with your study group, or reminders to yourself to go back and study a certain concept again to make sure that you really got it.

Having organized notes is essential at exam time and when doing homework assignments. Your ability to easily locate the important concepts of a recent lecture will help you move along more rapidly, as you don't have to spend time rereading an entire chapter just to reinforce one point that you may not have quite understood.

Your Guide is a valuable resource. You've found a wonderful study partner!

Note-Taking Tips

1. It is easier to take notes if you are not hearing the information for the first time. Read the chapter or the material that is about to be discussed before class. This will help you to anticipate what will be said in class, and have an idea of what to write down. It will also help to read over your notes from the last class. This way you can avoid spending the first few minutes of class trying to remember where you left off last time.
2. Don't waste your time trying to write down everything that your professor says. Instead, listen closely and only write down the important points. Review these important points after class to help remind you of related points that were made during the lecture.
3. If the class discussion takes a spontaneous turn, pay attention and participate in the discussion. Only take notes on the conclusions that are relevant to the lecture.
4. Emphasize main points in your notes. You may want to use a highlighter, special notation (asterisks, exclamation points), format (circle, underline), or placement on the page (indented, bulleted). You will find that when you try to recall these points, you will be able to actually picture them on the page.
5. Be sure to copy down word-for-word specific formulas, laws, and theories.
6. Hearing something repeated, stressed, or summed up can be a signal that it is an important concept to understand.
7. Organize handouts, study guides, and exams in your notebook along with your lecture notes. It may be helpful to use a three-ring binder, so that you can insert pages wherever you need to.
8. When taking notes, you might find it helpful to leave a wide margin on all four sides of the page. Doing this allows you to note names, dates, definitions, etc. for easy access and studying later. It may also be helpful to make notes of questions you want to ask your professor about or research later, ideas or relationships that you want explore more on your own, or concepts that you don't fully understand.
9. It is best to maintain a separate notebook for each class. Labeling and dating your notes can be helpful when you need to look up information from previous lectures.
10. Make your notes legible, and take notes directly in your notebook. Chances are you won't recopy them no matter how noble your intentions. Spend the time you would have spent recopying the notes studying them instead, drawing conclusions and making connections that you didn't have time for in class.
11. Look over your notes after class while the lecture is still fresh in your mind. Fix illegible items and clarify anything you don't understand. Do this again right before the next class.

Notes

Chapter 1

Life in the Balance

Fifth Edition Human Biology Daniel D. Chiras

Section 1-1 Health and Homeostasis

Homeostasis is a state of relative constancy.

- **Homeostasis** comes from two Greek words, homeo, which means "the same," and stasis, which means "standing.
- Homeostatic mechanisms exist at all levels of biological organization, from cells to organisms to ecosystems.
- These homeostatic mechanisms are vital to survival and reproduction.

Fifth Edition Human Biology Daniel D. Chiras — Section 1-1

Human health depends on maintaining healthy physical, psychological, and social environments.

- The health of the environment and the health of organisms, including human beings, are interdependent.
- adverse changes in the environment can affect on human health.
- The health of organisms also requires social and psychological conditions conducive to mental health.

Notes

Fifth Edition Human Biology Daniel D. Chiras — Section 1-1

Human health is a state of physical and mental well-being.

- Physical well-being: an absence of disease or symptoms of disease, a lack of risk factors that lead to disease, and good physical fitness.
- Mental health: a lack of mental illness and a capacity to deal effectively with the normal stresses and strains of life.

Fifth Edition Human Biology Daniel D. Chiras — Section 1-1

Health is dependent on properly functioning homeostatic mechanisms.

- When these mechanisms break down, illnesses often result.
- Stress results in disease by disrupting homeostatic mechanisms.
- not all diseases result from homeostatic imbalance
    - Genetics
    - Viruses
    - Bacteria

Fifth Edition Human Biology Daniel D. Chiras

Section 1-2 Evolution and the Characteristics of Life

All life-forms alive today exist because of evolution.

- The process of evolution results in improvements in existing species.
- It may also lead to the formation of new species.
- Evolution is responsible for the great diversity of life-forms.

Notes

Fifth Edition **Human Biology** Daniel D. Chiras — Section 1-2

- **monerans** were the first to evolve
- monerans gave rise to a more complex **protistans**
- protistans gave rise to three major groups, or kingdoms: plants, fungi, and animals

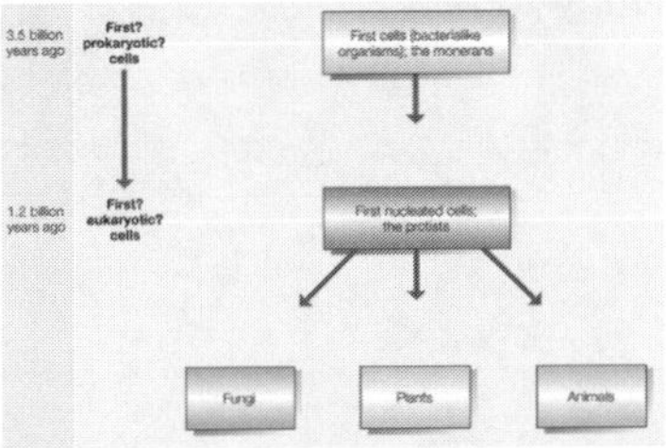

Figure1-5

Fifth Edition **Human Biology** Daniel D. Chiras — Section 1-2

Humans are similar to other organisms in many ways.

- are made up of cells
- grow and maintain their organization by taking in chemicals and energy
- exhibit metabolism
- rely on homeostatic mechanisms
- exhibit irritability
- are capable of reproduction and growth
- are products of evolutionary development and are subject to evolutionary change

Fifth Edition **Human Biology** Daniel D. Chiras — Section 1-2

Humans are one form of life on Earth and have many unique characteristics.

- our ability to acquire and use complex languages
- our culture
- our ability to plan for the future
- our ability to shape the environment

Notes

Section 1-3 Understanding Science

Science is both a systematic method of discovery and a body of information about the world around us.

Fifth Edition Human Biology Daniel D. Chiras Section 1-3

The **scientific method** generally starts with observations that lead to hypotheses and experiments to test them.

Hypothesis: a tentative explanation of the phenomenon

- ***Experiment:*** a procedure designed to test some idea

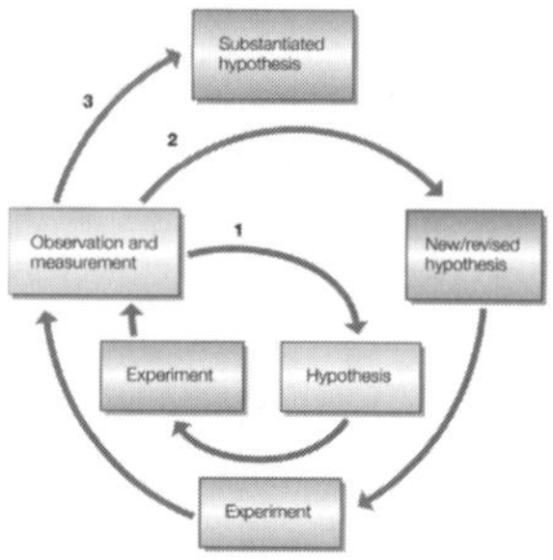

Figure1-8

Fifth Edition Human Biology Daniel D. Chiras Section 1-3

Theories are broad generalizations based on many experimental observations.

- Theories cannot be tested by single experiments because they encompass many bits of information.
- Theories can change over time as new information becomes available.

Fifth Edition Human Biology Daniel D. Chiras — Section 1-3

Science relies on inductive and deductive reasoning.

- **inductive reasoning**: a type of reasoning based on observations
- **deductive reasoning**: a specific statement or conclusion based on a general rule

Fifth Edition Human Biology Daniel D. Chiras — Section 1-3

Science helps shape our lives and our values.

- Science and the scientific process are essential to modern existence.
- Science can also influence political decisions.
- Many decisions in the public-policy arena are influenced by values rather than scientific facts.
- However,science can even influence human values.

Fifth Edition Human Biology Daniel D. Chiras

Section 1-4 Critical Thinking

Critical thinking rules allow us to carefully analyze problems, issues, and information for errors of reasoning.

Careful analysis helps us distinguish knowledge from beliefs or judgments.

Notes

Notes

Fifth Edition Human Biology Daniel D. Chiras — Section 1-4

Seven Rules for Critical Thinking

1. Gather complete information
2. Understand and define all terms
3. Question the methods by which data and information were derived
4. Question the conclusions
5. Uncover assumptions and biases
6. Question the source of the information
7. Understand your own biases and values

Notes

Fifth Edition Human Biology Daniel D. Chiras

2-1 Atoms and Subatomic Particles

Atoms are the fundamental unit of all matter.

- Atoms contain electrons, protons, and neutrons.

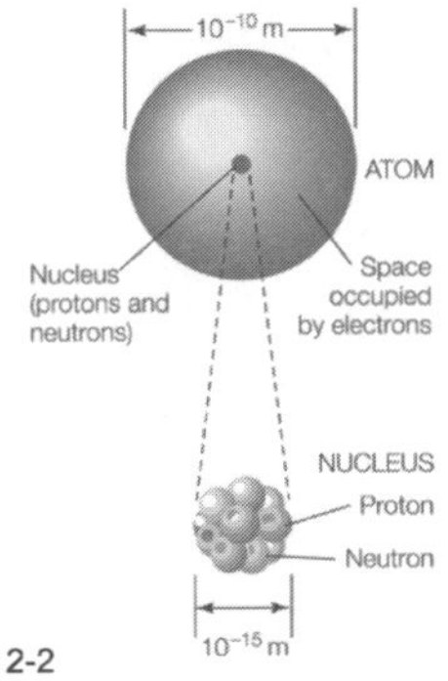

Figure 2-2

Fifth Edition Human Biology Daniel D. Chiras — Section 2-1

Elements are pure substance that contains only one type of atom.

- 92 naturally occurring elements are known
- only about 20 are found in organisms
- Four elements in this group: carbon, oxygen, hydrogen, and nitrogen (remember: COHN) comprise 98% of the atoms of all living things

Notes

Isotopes are alternative forms of atoms, differing only in the number of neutrons additional neutrons makes some atoms a bit unstable.

- To achieve a more stable state, many isotopes release tiny energetic particles from their nuclei.
- These emissions are known as radiation
- Radioactive isotopes are called **radionuclides**

Section 2-2 The Making of a Molecule

Atoms combine to form molecules.

- ***Compound***: a substance made up of two or more atoms
- ***Molecule***: the smallest particle of a compound that still retains the properties of that compound.

Atoms bond to form more stable configurations.

- There are two types of bonds that form between atoms:
  - Ionic
  - Covalent
- Electrons are responsible for creating the bonds that hold atoms together

Fifth Edition Human Biology Daniel D. Chiras — Section 2-2

Ionic bonds are electrostatic attractions between two oppositely charged particles.

- ***Ionic*** bonds form between two atoms when one loses an electron and the other gains an electron
- This reaction creates two charged particles, known as ***ions***

Fifth Edition Human Biology Daniel D. Chiras — Section 2-2

Covalent bonds are formed by the sharing of electrons between atoms

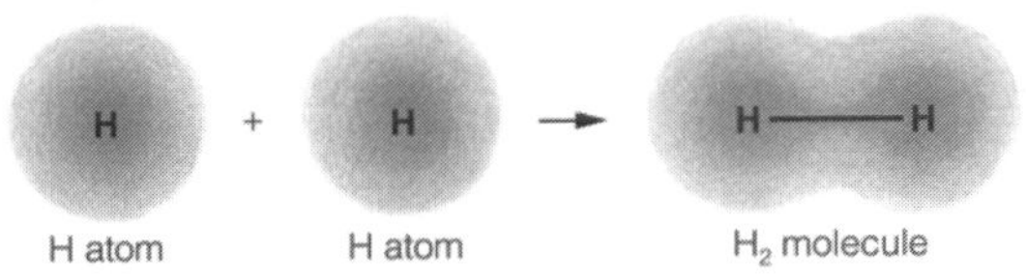

Figure 2-4

Fifth Edition Human Biology Daniel D. Chiras — Section 2-2

Polar covalent bonds occur any time there is an unequal sharing of electrons by two atoms

- A polar covalent bond's atoms bear a slight charge—either positive or negative

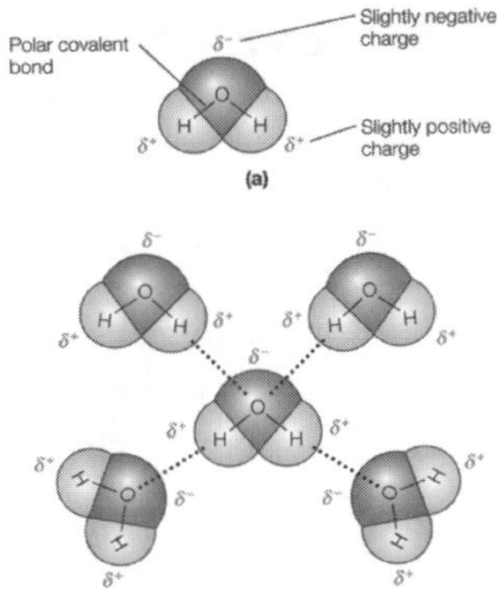

Figure 2-7

Notes

Notes

Hydrogen bonds form between slightly charged atoms usually on different molecules.

- In water, the electrostatic attractions between the positively charged hydrogen atoms of one water molecule and the negatively charged oxygen atoms form the hydrogen bond.

Chemical compounds fall into two broad groups: organic and inorganic.

- ***Organic compounds*** contain molecules that are made primarily of carbon atoms.
- ***Inorganic compounds*** are compounds that are not organic

Section 2-3 Water, Acids, Bases, and Buffers

Water is vital to life for many reasons.

- Water is a major component of all cells and organisms
- Water serves as a solvent, a transport medium, and a lubricant.
- Water participates in many chemical reactions.
- Water helps regulate body temperature.

Notes

Water molecules split into hydrogen and hydroxide ions. This reaction is reversible

- The ratio of water molecules to the ions, H^+ and OH^-, in the human body is about 500 million: 1
- The slightest change in the hydrogen ion concentration can alter cells and organisms, shutting down biochemical pathways
- Humans contain several homeostatic mechanisms to ensure a constant level of these ions

Fifth Edition Human Biology Daniel D. Chiras Section 2-3

Acids are substances that add hydrogen ions to solution; bases remove them.

- Acidity is measured on the ***pH scale***
- A solution with a pH less than 7 is acidic.
- A solution with a pH greater than 7 is basic.
- On the pH scale, a change in one pH unit represents a tenfold change in acidity
- Most biochemical reactions occur at pH values between 6 and 8.

Fifth Edition Human Biology Daniel D. Chiras Section 2-3

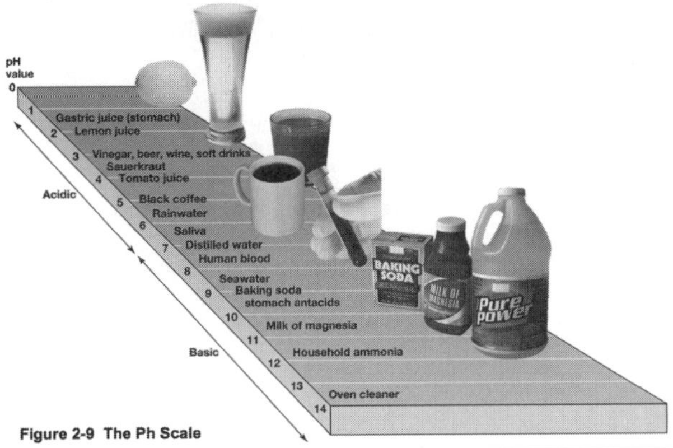

Figure 2-9 The Ph Scale

Notes

Homeostasis is ensured in part by ***buffers***, molecules that help maintain pH within a narrow range.

- Buffers help maintain a constant pH by removing hydrogen ions from solution when levels increase.
- Buffers give back the hydrogen ions when levels fall.

2-4 Overview of Other Biologically Important Molecules

Carbohydrates, lipids, amino acids, proteins, and nucleic acids are biologically important molecules.

Notes

Fifth Edition Human Biology Daniel D. Chiras

Section 3-1 Microscopes
Illuminating the Structure of Cells

Microscopes fall into two broad categories:

- light microscopes
  - uses ordinary visible light to illuminate the specimen
  - can magnify objects from 100 to 400 times their original size
- electron microscopes
  - uses a beam of electrons to create a visual image of the specimen
  - can enlarge objects 100,000 times their original size

Fifth Edition Human Biology Daniel D. Chiras

Section 3-2 An Overview of Cell Structure

The cell consists of two main compartments.

- The nuclear (nucleus) compartment
- The cytoplasmic compartment

Notes

Fifth Edition Human Biology Daniel D. Chiras **Section 3-2**

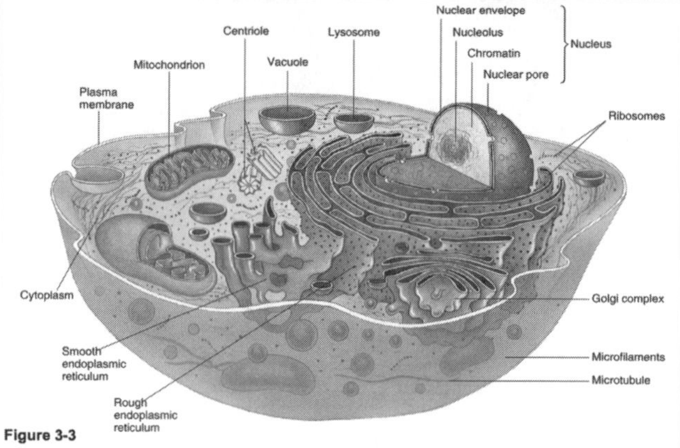

Figure 3-3

Fifth Edition Human Biology Daniel D. Chiras **Section 3-2**

- The nuclear (nucleus) compartment
  - The nucleus contains the genetic information that regulates the structure and function of all eukaryotic cells.
  - The **cytoplasm** contains numerous cellular **organelles,** which perform specific functions.
- The cytoplasmic compartment
  - The cytoplasmic compartment is home to the **cytoskeleton**.
    - The cytoskeleton is a network of tubules and filaments.
    - The cytoskeleton also binds to enzymes in metabolic

Fifth Edition Human Biology Daniel D. Chiras

Section 3-3 The Structure and Function of the Plasma Membrane

The **plasma membrane** is the outermost boundary of the cell.

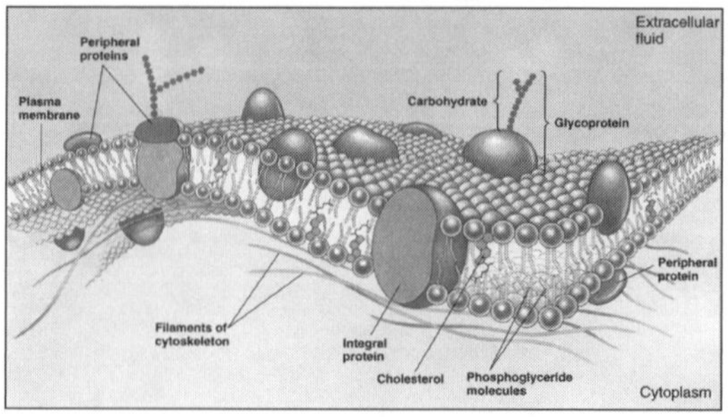

Figure 3-6

Notes

Fifth Edition Human Biology Daniel D. Chiras — Section 3-3

The plasma membrane consists of lipids, protein, and carbohydrate.

- **Lipids** (fats) are biochemicals characterized by their lack of water solubility.
- Human cells have thousands of different types of **proteins**, each type structurally and functionally unique.
- **Amino acids** are small organic molecules that join by peptide bonds, forming peptides and proteins.
- The **lipids** of the plasma membrane form a double layer in which many of the proteins float freely.

Fifth Edition Human Biology Daniel D. Chiras — Section 3-3

The plasma membrane is essential to cellular homeostasis.

- The plasma membrane regulates the flow of molecules and ions into and out of the cell.
- The plasma membrane is selectively permeable

Fifth Edition Human Biology Daniel D. Chiras — Section 3-3

Molecules move through the plasma membrane in five ways.

- **Diffusion**
- **Carrier Proteins and Diffusion**
- **Active Transport** --molecules are also actively transported across the membrane
- **Endocytosis**
- **Exocytosis**

Notes

Fifth Edition Human Biology Daniel D. Chiras — Section 3-3

Diffusion: The movement of molecules from high to low concentrations.

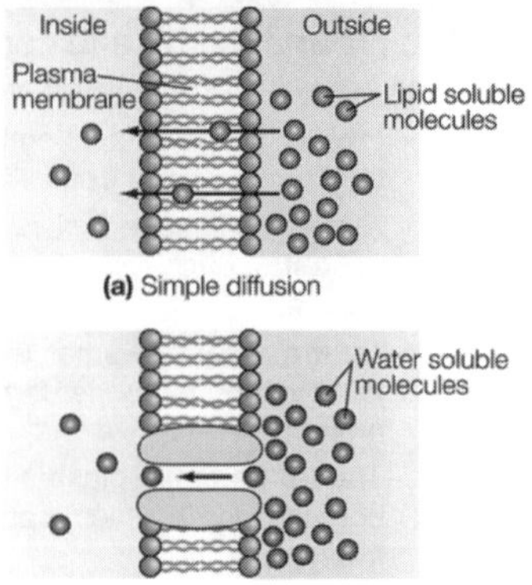

(a) Simple diffusion

(b) Diffusion through protein pores

Figure 3-12

Fifth Edition Human Biology Daniel D. Chiras — Section 3-3

Carrier Proteins and Diffusion: Water-soluble molecules also diffuse through membranes with the aid of carrier molecules.

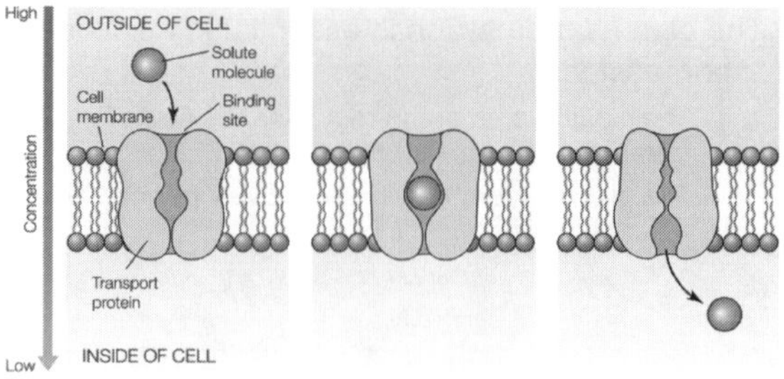

(a) Facilitated diffusion

Figure 3-13

Fifth Edition Human Biology Daniel D. Chiras — Section 3-3

Active Transport: Molecules are also actively transported across the membrane.

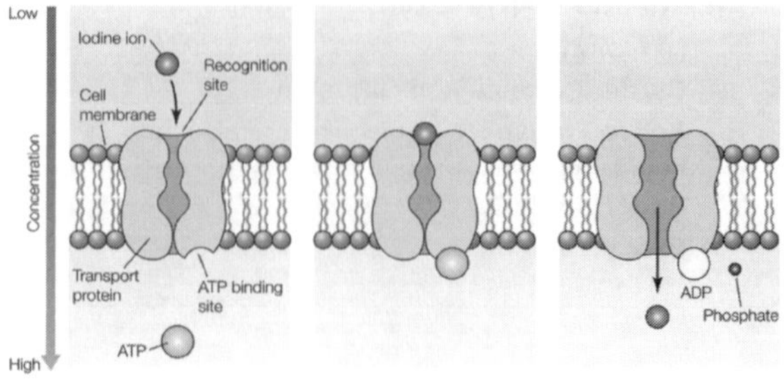

(b) Active transport

Figure 3-13

Fifth Edition Human Biology Daniel D. Chiras **Section 3-3**

Active Transport: Molecules are also actively transported across the membrane.

- **ATP** consists of 3 smaller molecules: adenine, ribose; and three phosphate groups.
- When the cell needs energy ATP splits off a phosphate, forming ADP
- The breaking of the phosphate bond yields energy that the cell can use directly
  - ATP --→ ADP + P_i + Energy
  - Reactions that give off energy often turn the energy over to ADP and new ATP can be formed.
  - ADP + P_i --→ Energy + ATP

Fifth Edition Human Biology Daniel D. Chiras **Section 3-3**

Endocytosis: Large molecules and cells are ingested by endocytosis.

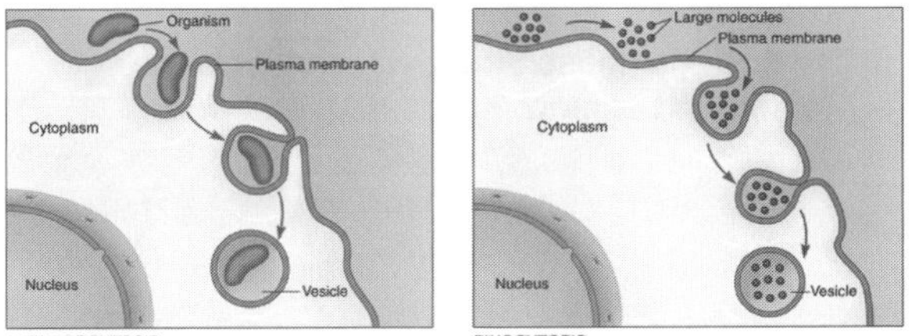

(a) PHAGOCYTOSIS PINOCYTOSIS

Figure 3-15

Fifth Edition Human Biology Daniel D. Chiras **Section 3-3**

Exocytosis: Cells also regurgitate materials, releasing large molecules.

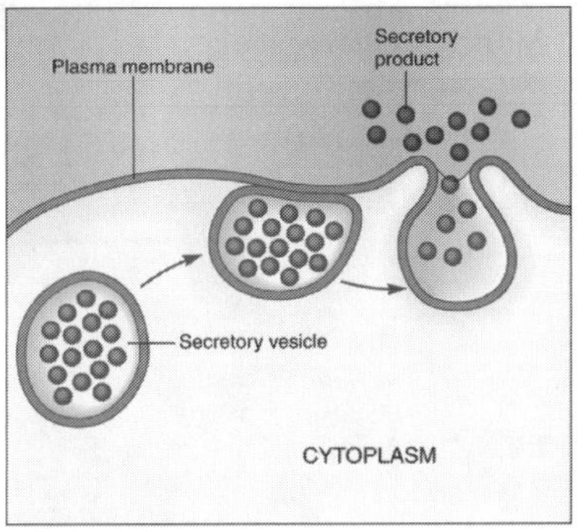

(b) EXOCYTOSIS

Figure 3-15

Notes

Notes

Fifth Edition Human Biology Daniel D. Chiras — Section 3-3

Osmosis: The diffusion of water across the plasma membrane is known as osmosis

- Whenever two fluids with different concentrations of solute are separated by a selectively permeable membrane, the fluid will flow from one to the other, moving down the concentration gradient.
- In humans, osmosis helps regulate the concentration of fluid surrounding the cell.

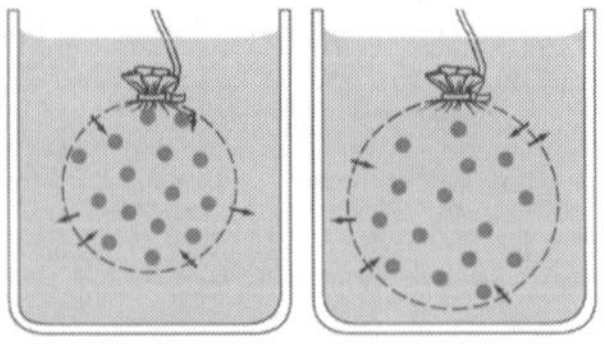

Figure 3-16

Fifth Edition Human Biology Daniel D. Chiras

Section 3-4 Cellular Compartmentalization: Organelles

The nucleus houses the DNA and is the cell's command center.

Fifth Edition Human Biology Daniel D. Chiras — Section 3-4

Mitochondria are the site of cellular energy production.

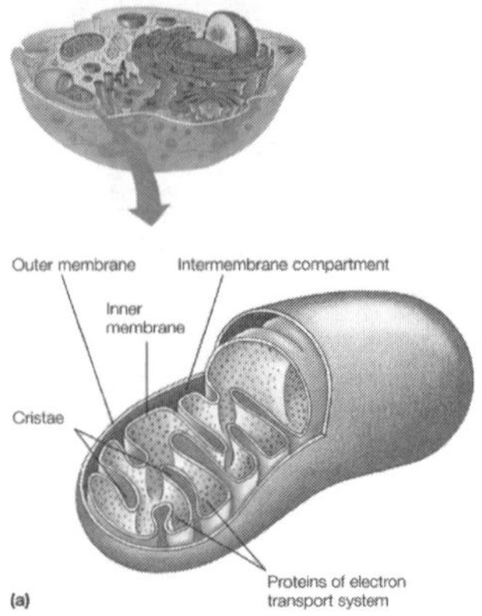

Figure 3-19

Fifth Edition Human Biology Daniel D. Chiras — Section 3-4

- Glucose breakdown takes place mostly in the mitochondrion.
- Energy liberated from glucose molecules is captured by ATP molecules.
- Cellular energy production captures 30–60% of the energy contained in glucose; the rest is given off as heat.

Fifth Edition Human Biology Daniel D. Chiras — Section 3-4

Three organelles are involved in manufacturing **protein**

- **Endoplasmic reticulum**
  - Some proteins are produced on the surface of the rough endoplasmic reticulum
- **Ribosomes**
  - Ribosomes are involved in all protein synthesis.
- **The Golgi complex**
  - The golgi modifies and packages proteins that are secreted from the cell for extracelluar use

Fifth Edition Human Biology Daniel D. Chiras — Section 3-4

Lysosomes are membrane-bound organelles that contain digestive enzymes

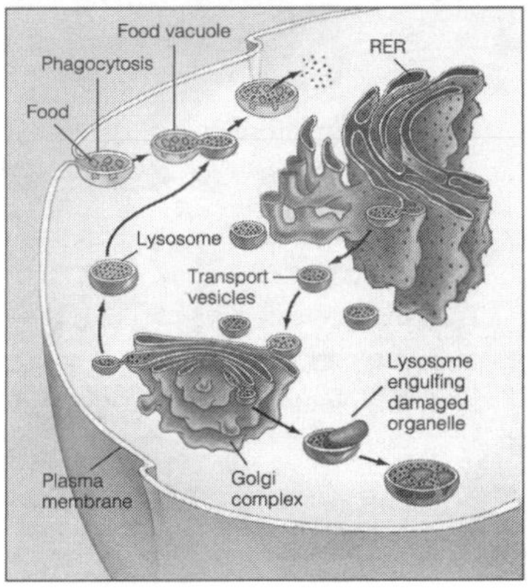

Figure 3-22a

Notes

Notes

Flagella are organelles that permit cellular motility.

Movement across the surface of cells is provided by cilia.

Some cells move by amoeboid motion.

Section 3-5 Energy and Metabolism

Cells get the energy they need primarily from two sources:
- the breakdown of glucose (a type of carbohydrate)
- the breakdown of triglycerides (a type of fat)

The chemical breakdown of glucose occurs in four steps.

1. Glycolysis
2. the transition reaction
3. the citric acid cycle
4. the electron transport system

Fifth Edition Human Biology Daniel D. Chiras — Section 3-5

- **1. Glycolysis** is the breakdown of glucose into two pyruvate molecules
  - Glycolysis nets the cell two molecules of ATP and two NADH molecules containing high-energy electrons.

Fifth Edition Human Biology Daniel D. Chiras — Section 3-5

- **2. The Transition Reaction**
  - In the transition reaction one carbon atom is cleaved off each pyruvate.
  - The transition reaction—pyruvate produced during glycolysis diffuses into the mitochondrion and undergoes the transition reaction.
  - This step produces acetyl CoA and two NADH molecules.

Fifth Edition Human Biology Daniel D. Chiras — Section 3-5

- **3. The citric acid cycle**
  - The citric acid cycle completes the breakdown of glucose.
  - The Acetyl CoA enters the citric acid cycle, which produces two ATPs and numerous molecules of NADH and FADH2.
  - These electron carriers accept high-energy electrons from various reactions in the citric acid cycle and move them to the electron transport system.

Notes

Notes

- **4. The Electron Transport System**
  - Electrons given off during the citric acid cycle and glycolysis enter the electron transport system.
  - The electron transport system is a series of protein molecules embedded in the inner surface of the inner membrane of the mitochondrion.

Enzymes are essential to most chemical reactions in cells.

- A missing enzyme in a metabolic pathway will shut the pathway down
- Each enzyme is a large protein with a small, uniquely shaped, region known as the **active site** where the chemical reaction occurs
- Enzymes are unchanged by the reaction and can be used over and over again.
- Enzymes can also be controlled by the cells themselves
- The control regions are called **allosteric sites**

Section 3-6 Fermentation

- Without oxygen, the electron transport system shuts down, as does the citric acid cycle.
- Cells must rely on glycolysis and fermentation to generate energy
- Fermentation is a chemical reaction in which pyruvate is converted into lactic acid or, depending on the organism, ethyl alcohol
- Lactic acid fermentation is an important source of food for humans

Notes

Fifth Edition Human Biology Daniel D. Chiras

Section 4-1 From Cells to Organ Systems

Cells unite to form tissues, and tissues combine to form organs.

Cells combine to form four primary tissues.

- Epithelial tissue
- Connective tissue
- Muscle tissue
- Nervous tissue

Fifth Edition Human Biology Daniel D. Chiras — Section 4-1

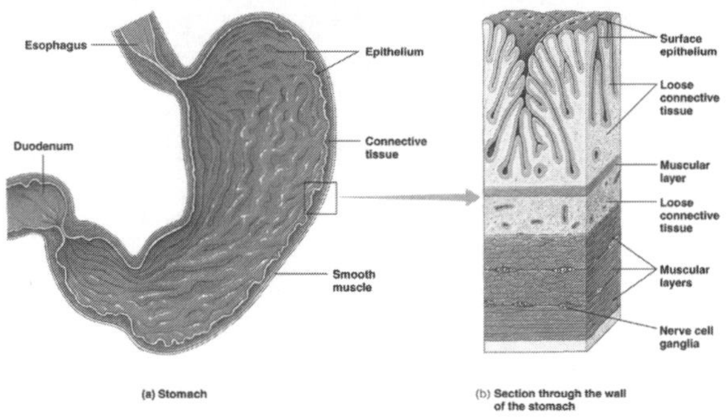

Figure 4-1 Human Stomach

Notes

Fifth Edition Human Biology Daniel D. Chiras **Section 4-1**

Epithelium forms the lining or external covering of organs and also forms glands.

- Epithelial tissues consist of two types:
  - membranous epithelia, which form the coverings or linings of organs,
  - glandular epithelia, which form exocrine and endocrine glands.

Epithelial tissues illustrate a basic biological principle: that structure closely correlates with function.

Fifth Edition Human Biology Daniel D. Chiras **Section 4-1**

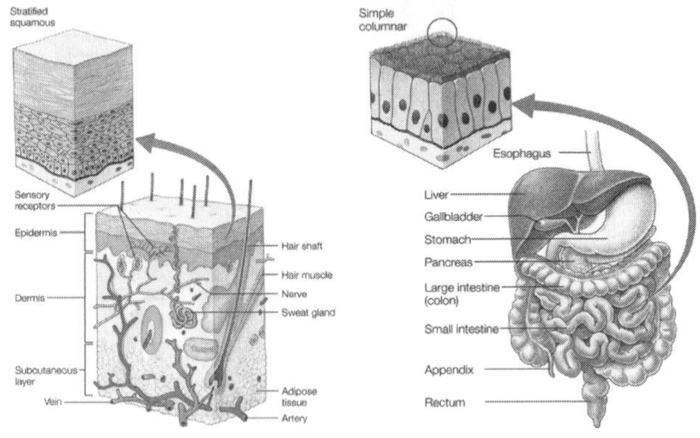

Figure 4-4

Fifth Edition Human Biology Daniel D. Chiras **Section 4-1**

Connective tissue binds the cells and organs of the body together.

- All connective tissues consist of two basic components: cells and extracellular fibers.
- Two types of connective tissue are found in the body:
  - Connective tissue proper
  - Specialized connective tissue.

Fifth Edition
Human Biology Daniel D. Chiras

Section 4-1

Connective tissue proper is an important structural component of organs.

- Connective tissue proper consists of two types:
  - dense connective tissue
  - loose connective tissue

Fifth Edition
Human Biology Daniel D. Chiras

Section 4-1

Specialized connective tissues perform specific functions essential to homeostasis

- The body contains three types of specialized connective tissue:
  - Cartilage
  - Bone
  - Blood

Fifth Edition
Human Biology Daniel D. Chiras

Section 4-1

–**Cartilage** consists of specialized cells embedded in a matrix of extracellular fibers and other extracellular material.

–**Bone** consists of bone cells (osteocytes) and a calcified cartilage matrix.

•Two types of bone tissue exist: spongy and compact.

–**Blood** contains blood cells, platelets, plasma.

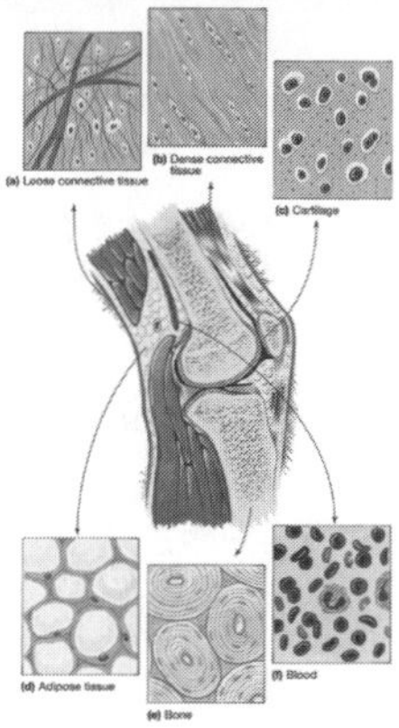

Figure 4-5

Notes

Notes

Fifth Edition Human Biology Daniel D. Chiras — Section 4-1

Muscle tissue consists of specialized cells that contract when stimulated.

- The body has three types of muscle tissue:
  - **Skeletal** (voluntary)
  - **Cardiac** (involuntary)
  - **Smooth muscle** (involuntary)

Fifth Edition Human Biology Daniel D. Chiras — Section 4-1

Nervous tissue contains specialized cells that conduct impulses

–Conducting cells, called **neurons**, transmit impulses from one region of the body to another.

–Nonconducting cells, **neuroglia,** are a type of nervous system connective tissue.

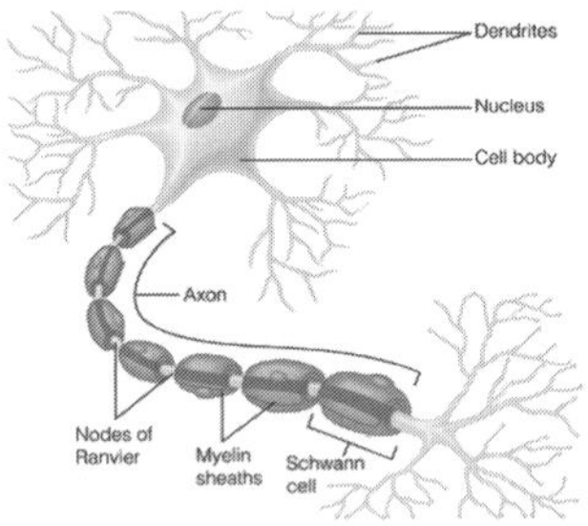

Figure 4-12c

Fifth Edition Human Biology Daniel D. Chiras — Section 4-1

Tissues combine to form organs.

- **Organs** are discrete structures that have evolved to perform specific functions

- Most organs do not function alone, they are part of a group of cooperative organs, called an **organ system**.

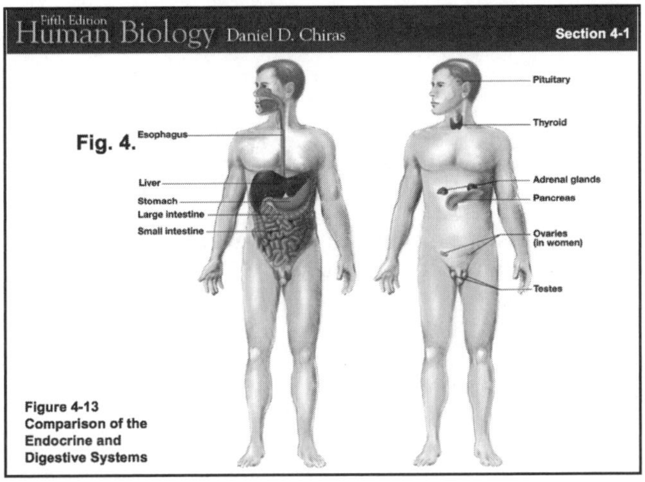

Fifth Edition Human Biology Daniel D. Chiras

Section 4-2 Principles of Homeostasis

Homeostatic systems operate primarily through **negative feedback** mechanisms.

Homeostatic feedback mechanisms require a **sensor** and an **effector**.

Homeostasis can be upset by changes in input and output.

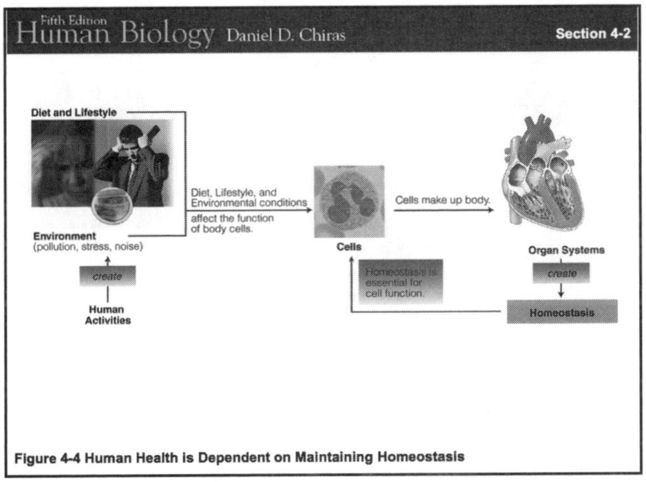

Notes

Notes

Homeostatic control requires the action of nerves and hormones.

Homeostatic mechanisms are reflexes, occurring subconsciously.

- Two types of reflexes exist: nervous system and hormonal.
- Nervous and endocrine mechanisms generally occur over considerable distances.

Section 4-3 Biological Rhythms

Physiological processes that fluctuate normally do so in predictable ways.

- **Circadian rhythms** are natural body rhythms linked to the 24-hour day-night cycle.

Internal biological rhythms are controlled by the brain.

Section 4-4 Health and Homeostasis

Human health depends on homeostasis.

- Homeostasis, in turn, requires
  - proper diet and exercise
  - freedom from disease-causing organisms
  - manageable levels of stress
  - a clean environment

Chapter 5: Nutrition and Digestion

Notes

Chapter 5

Nutrition and Digestion

Fifth Edition Human Biology Daniel D. Chiras

Section 5-1 An Introduction to Nutrition

A healthy diet consists of:

- Breads, cereals, rice, and pasta.
- Plenty of fruits and vegetables
- Meat and milk products. but in lesser amounts.
- Minimal fat

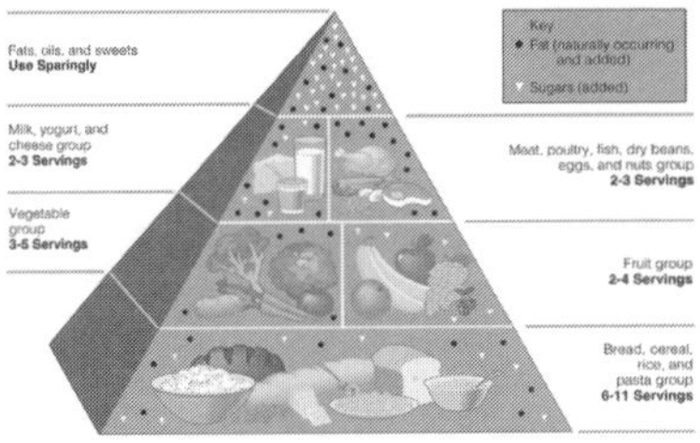

Figure 5-1

Fifth Edition Human Biology Daniel D. Chiras — Section 5-1

Macronutrients are needed in large quantity and include **water, carbohydrates, lipids**, and **proteins**.

Water
- is involved in many chemical reactions in the body.
- helps maintain body temperature and a constant level of nutrients and wastes in body fluids.

Carbohydrates
- are a major source energy
- belong to three groups:
  - monosaccharides, disaccharides, and polysaccharides

Notes

Fifth Edition
Human Biology Daniel D. Chiras
Section 5-1

Lipids

- are a source of energy but also play a structural role
  - Triglycerides (fats and oil) provide energy and insulation.
  - Phospholipids and steroids are part of the plasma membranes of cells.
- Triglycerides with many double bonds in their fatty acid side chains (the polyunsaturated fatty acids) lower one's risk of developing atherosclerosis.
- Triglycerides low in saturated fatty acids, found in animal fats, tend to increase atherosclerosis.

Fifth Edition
Human Biology Daniel D. Chiras
Section 5-1

Amino Acids and Protein.

- Proteins in food are broken down to amino acids which are used to make proteins such as enzymes and hormones.
- Amino acids the body cannot synthesize are known as **essential amino acids**.
- Amino acids the body can synthesize are known as nonessential amino acids.

Fifth Edition
Human Biology Daniel D. Chiras
Section 5-1

Overnutrition is now reaching epidemic proportions in the more developed countries of the world.

- Many people ingest nutrients in quantities far greater than is necessary.
- The effects of overeating include heart attacks, stroke, late-onset diabetes, and osteoarthritis.

Fifth Edition Human Biology Daniel D. Chiras — Section 5-1

Micronutrients are substances required in minute quantities and include vitamins and minerals

- Vitamins play a key role in many chemical reactions in the body
- Vitamins fall into two broad categories: **water-soluble** and **fat-soluble**
- Vitamin deficiencies deficiencies can lead to serious homeostatic imbalance and many adverse health effects.

Fifth Edition Human Biology Daniel D. Chiras — Section 5-1

- Minerals fit into one of two groups:
  - **trace** minerals, those required in very small quantity (< 5 grams)
  - **major minerals**, those required in greater quantity (> 5 grams).

Fifth Edition Human Biology Daniel D. Chiras — Section 5-1

Foods contains substances that reduce cholesterol and fight cancer

- **Antioxidants** react with chemical substances in the blood that are involved in cholesterol build-up in artery walls.
- **Flavonoids** reduce plaque buildup and help stimulate the relaxation of smooth muscle in arteries.

Notes

Notes

Fifth Edition Human Biology Daniel D. Chiras

Section 5-2: The Digestive System

Physical breakdown of food occurs in the mouth
- Chewing breaks food into smaller pieces
- **Saliva** performs at least five functions:
  - liquefies the food
  - kills or neutralizes some bacteria
  - dissolves substances so they can be tasted
  - begins to break down starch molecules
  - cleanses the teeth

Fifth Edition Human Biology Daniel D. Chiras — Section 5-2

Teeth must be kept clean to prevent plaque and cavities.
- Each tooth consists of:
  - *enamel*—the hard outer layer
  - *dentin*—a softer layer
  - pulp—the inner region filled with nerves and blood vessels
- Bacteria that live on the teeth secrete a sticky, acidic material called **plaque.**

Fifth Edition Human Biology Daniel D. Chiras — Section 5-2

The tongue plays a key role in swallowing.
- The tongue pushes food to the back of the oral cavity into the **pharynx.**
- The tongue contains taste receptors on its upper surface.

Fifth Edition Human Biology Daniel D. Chiras — Section 5-2

The epiglottis keeps food from entering the trachea.
- Food entering the pharynx triggers the swallowing reflex, which forces the food into the esophagus.

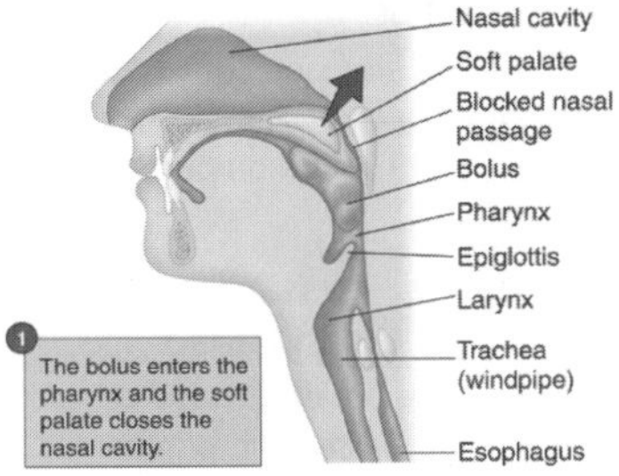

Figure 5-15 (1)

Fifth Edition Human Biology Daniel D. Chiras — Section 5-2

The esophagus transports food to the stomach via peristalsis

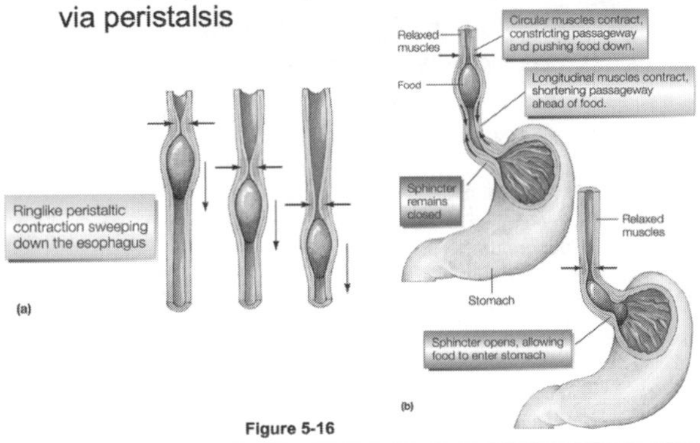

Figure 5-16

Fifth Edition Human Biology Daniel D. Chiras — Section 5-2

The stomach stores food, releasing it into the small intestine in spurts.
- Gastric glands in the stomach produce
  - hydrochloric acid
  - Pepsinogen
- The food is turned into chyme before it is released.

Notes

Notes

Fifth Edition Human Biology Daniel D. Chiras — Section 5-2

Chyme leaves the stomach and enters the small intestine.
- The stomach contents are emptied in 2–6 hours.
- As chyme and protein leave the stomach, gastric gland secretion declines.

The small intestine serves as a site of food digestion and absorption.
- The small intestine consists of three parts:
  - the duodenum, jejunum, and ileum.

Fifth Edition Human Biology Daniel D. Chiras — Section 5-2

Digestion in the small intestine requires enzymes from two sources.
- the lining of the intestine
- the **pancreas**
  - produces enzymes and sodium bicarbonate essential for the digestion.
  - The digestive enzymes of the pancreas flow into the duodenum.

Fifth Edition Human Biology Daniel D. Chiras — Section 5-2

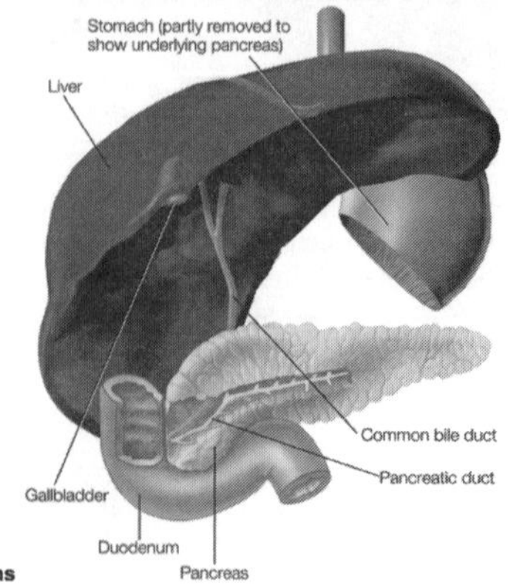

Figure 5-18 The Organs of Digestion

Fifth Edition Human Biology Daniel D. Chiras — Section 5-2

The liver carries out hundreds of important functions.
- The liver stores minerals, vitamins, glucose, and lipids
- The liver produces an emulsifying agent, **bile**, which plays a key role in the digestion of fat.
- Bile is transported to the **gallbladder** until it is needed

Fifth Edition Human Biology Daniel D. Chiras — Section 5-2

The intestinal epithelium, the lining of the intestine, is specially modified for absorption.
- The lining of the small intestine is folded.
- On the surfaces of the folds are fingerlike projections known as **villi** and **microvilli.**

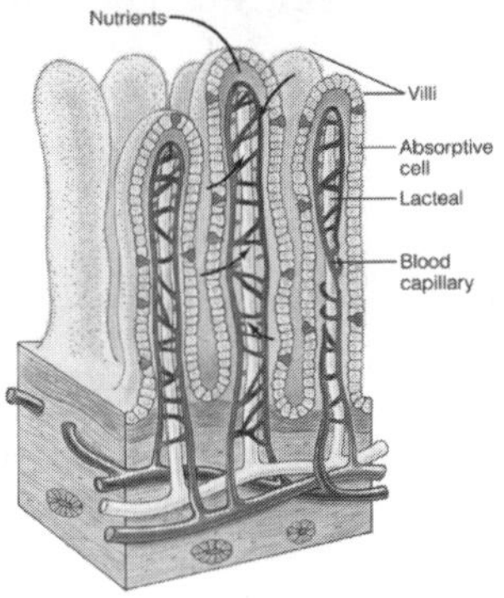

Figure 5-19e

Fifth Edition Human Biology Daniel D. Chiras — Section 5-2

The large intestine is the site of water resorption.
- It consists of four regions:
  - the cecum, appendix, colon, and rectum
- It also absorbs sodium and potassium as well as vitamins produced by intestinal bacteria.
- It transports the waste, or feces, to the outside of the body.

Notes

Notes

Fifth Edition Human Biology Daniel D. Chiras

Section 5-2

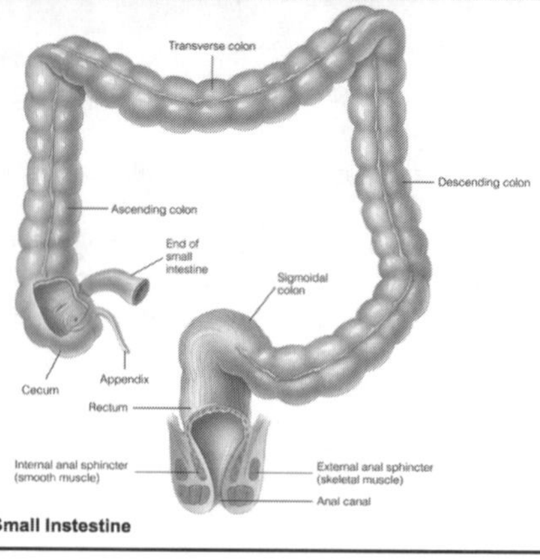

Figure 5-19 The Small Instestine

Fifth Edition Human Biology Daniel D. Chiras

Section 5:3 Controlling Digestion

Digestive processes are largely controlled by the nervous and endocrine systems.

Salivation is stimulated by a nervous reflex.

Gastric gland secretion is controlled by a variety of different mechanisms.

- External stimuli
- Protein in the stomach

Fifth Edition Human Biology Daniel D. Chiras

Section 5-3

Pancreatic secretions are stimulated by two intestinal hormones

- **secretin** produced by the cells duodenum and stimulated by the presence of acidic chyme in the small intestine.
- **cholecystokinin** produced by cells of the duodenum in the presence of chyme.

Notes

Fifth Edition Human Biology Daniel D. Chiras

Section 6-1 The Circulatory System's Function
An Overview

The circulatory system consists of a pump, the heart, and a network of vessels.

- The circulatory system is one of the body's chief homeostatic systems. The system:
  - Distributes oxygen
  - helps maintain constant levels of nutrients and wastes,
  - helps regulate body temperature,
  - distributes body heat,
  - protects against microorganisms, and
  - through clotting, protects against blood loss.

Fifth Edition Human Biology Daniel D. Chiras

Section 6-2 The Heart

The heart is a muscular organ that pumps blood throughout the body

- There are three layers in the heart walls:
  - The pericardium
  - The myocardium
  - The endocardium

Notes

Fifth Edition Human Biology Daniel D. Chiras — Section 6-2

Blood flows through two distinct circuits in the cardiovascular system.

- The **pulmonary circuit**, which transports blood to and from the lungs.
- The **systemic circuit**, which delivers blood to the body and returns it to the heart.

Fifth Edition Human Biology Daniel D. Chiras — Section 6-2

The human heart consists of four chambers: two atria and two ventricles.

- The right atrium and right ventricle service the pulmonary circuit.
- The right atrium receives blood, low in oxygen and rich in carbon dioxide, from the **superior** and **inferior vena cavae.**
- From the right atrium, blood is pumped into the right ventricle, then to the lungs.

Fifth Edition Human Biology Daniel D. Chiras — Section 6-2

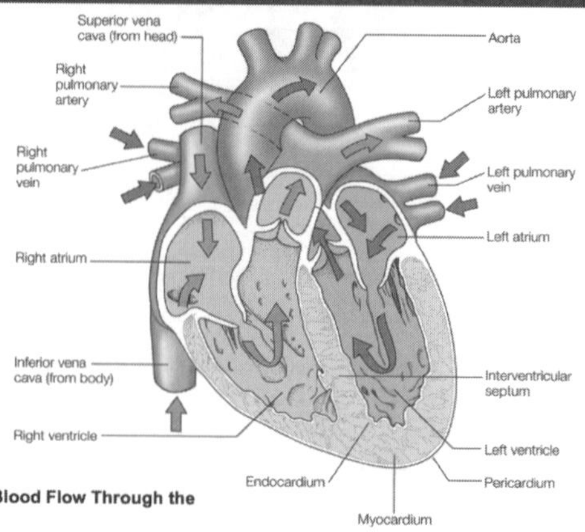

Figure 6-2 Blood Flow Through the Heart

- Blood, resupplied with oxygen, returns to the heart via the pulmonary veins, which empty into the left atrium.
- The left atrium and left ventricle pump blood into the **aorta** and are part of the systemic circuit.
- Blood from the aorta supplies cells of tissues and organs with oxygen and picks up cellular wastes.

Fifth Edition Human Biology Daniel D. Chiras — Section 6-2

Valves help control the flow of blood in the heart

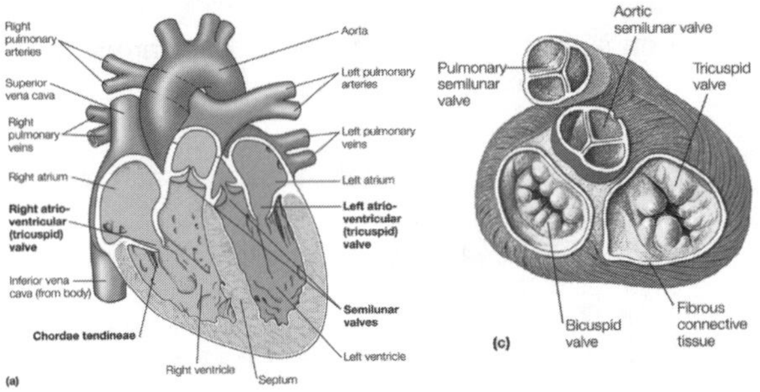

Heart sounds result from the closing of heart valves

Fifth Edition Human Biology Daniel D. Chiras — Section 6-2

Heart rate is controlled primarily by an internal pacemaker .

- The **sinoatrial (SA) node** is located in the upper wall of the right atrium.
- The cells of the SA node discharge periodically, sending impulses to all atrial muscle cells, which cause them to contract in unison.
- The impulse next travels to the ventricles, but its passage is delayed to time for the ventricles to fill.
- The impulse is channeled the **atrioventricular (AV) node**.

Notes

Notes

Fifth Edition Human Biology Daniel D. Chiras — Section 6-2

External factors regulate the pacemaker

- The SA node is curbed by impulses transmitted by nerves that connect the heart with a control center in the brain.
- Hormones also play a role in controlling heart rate.

The heart's electrical activity can be measured on the surface of the chest.

- The measurement is called an **electrocardiogram.**

Fifth Edition Human Biology Daniel D. Chiras

Section 6-3 The Blood Vessels

Blood flows from the heart to the capillaries, then back to the heart via veins.

Arteries and arterioles deliver oxygen-rich blood to tissues and organs.

- The **aorta** and many of its chief branches are elastic.
- **Blood pressure** and flow rate are highest in the aorta and drop considerably as the arteries branch.
- This decline enhances the rate of exchange between the blood and the tissues.

Fifth Edition Human Biology Daniel D. Chiras — Section 6-3

Capillaries permit the exchange of nutrients and wastes

- Capillaries form branching networks, known as capillary beds, among the cells of body tissues

Venules drain capillaries and merge to form veins that transport blood back to the heart.

- In the veins above the heart, gravity is the chief means of propulsion.
- Veins below the heart rely on the movement of body parts to squeeze the blood upward and on valves that prevent the backflow of blood

Fifth Edition Human Biology Daniel D. Chiras

Section 6-4 The Lymphatic System

The lymphatic system is a secondary system of vessels that returns excess fluid to the circulatory system

- The lymphatic system includes the lymph nodes, spleen, thymus, and tonsils.
- Lymph nodes filter the lymph.
- Normally, lymph is removed from tissues at a rate equal to its production.

Fifth Edition Human Biology Daniel D. Chiras — Section 6-4

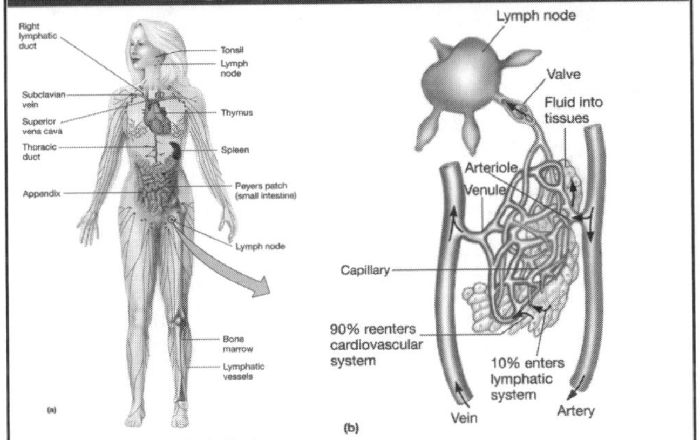

Figure 6-14 The Lymphatic System

Fifth Edition Human Biology Daniel D. Chiras

Section 6-5 Cardiovascular Diseases
Causes and Cures

The cardiovascular system—the heart and blood vessels—are subject to a variety of potentially life threatening disease.

Notes

Notes

Myocardial infarction: blockage of the arteries supplying heart muscle can lead to heart attacks.

- can occur without warning or may be preceded by several weeks of **angina.**

Fibrillation: another type of heart attack results from a loss of electrical control.

- The cardiac muscle cells beat independently.
- The heart can stop beating altogether (**cardiac arrest**).

Preventing heart attacks is the best cure

- Proper diet, exercise, and stress management can reduce the risk of heart problem

Many medical treatments are available to treat heart attacks.

High blood pressure (hypertension) develops gradually over time.

- **Hypertension** is a prolonged elevation in blood pressure.
- Early detection and treatment are essential to prevent serious problems, including heart attacks.

Fifth Edition
Human Biology Daniel D. Chiras

Section 6-5

Atherosclerosis results from the buildup of cholesterol plaque in arteries.

- Arteries clogged with cholesterol force the heart to work harder.
- Blood clots can lodge in narrowed coronary arteries.

Weakening of the arterial walls can result in a rupture (aneurysm) of the wall.

Notes

Chapter 7: The Blood

Notes

Fifth Edition Human Biology Daniel D. Chiras

Section 7-1 Blood: An Overview

Blood consists of plasma and formed elements, mostly blood cells and platelets.

- The blood plasma is a watery transport medium.
- Blood consists of two basic components:
  - The **plasma**—fluid that contains dissolved nutrients, proteins, gases, and wastes.
  - The formed elements—red blood cells, white blood cells, and platelets, which are suspended in the plasma.
- Plasma constitutes about 55% of the blood volume, and the formed elements make up the remainder.

Fifth Edition Human Biology Daniel D. Chiras — Section 7-1

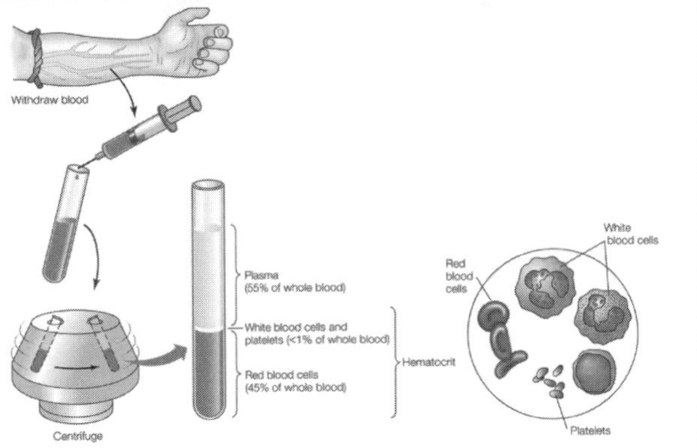

Figure 7-1 Blood Composition

Notes

Fifth Edition
Human Biology Daniel D. Chiras

Section 7-2 Blood Plasma

Plasma is a light yellow (straw-colored) fluid that plays many important functions in maintaining homeostasis.

- Plasma transports many important substances.
- Plasma proteins help regulate blood pH and osmotic pressure.
- Some plasma proteins serve as carrier proteins.

Fifth Edition
Human Biology Daniel D. Chiras

Section 7-3 Red Blood Cells

Red blood cells are highly flexible cells that transport oxygen in the blood.

- Red blood cells (RBCs) also transport carbon dioxide.

Red blood cells are replenished by stem cells in bone marrow.

- RBCs lack nuclei and organelles.
- RBCs are produced by stem cells.

Fifth Edition
Human Biology Daniel D. Chiras Section 7-3

Hemoglobin is an oxygen-transporting protein found in RBCs.

- Inside the RBC, oxygen binds to the iron in hemoglobin molecules for transport.

Anemia results in a decrease in oxygen transport by the blood.

Human Biology Fifth Edition Daniel D. Chiras

Section 7-4 White Blood Cells

White blood cells are a diverse group that protects the body from infection.

- White blood cells (WBCs) are produced in the bone marrow and circulate in the bloodstream.

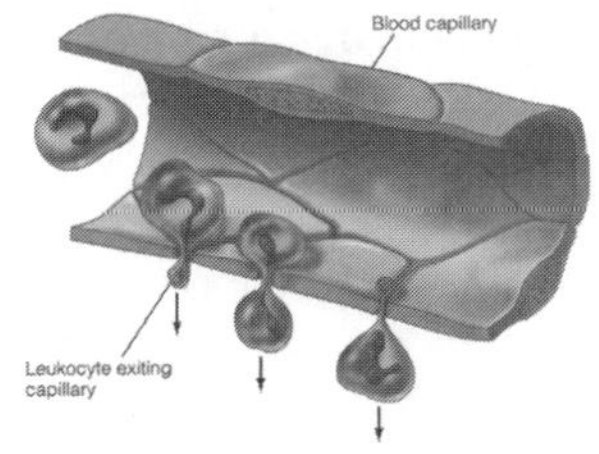

Figure 7-5

Human Biology Fifth Edition Daniel D. Chiras — Section 7-4

Neutrophils are phagocytic cells and are the first to arrive at an injury.

- Neutrophils leave the bloodstream and migrate to the site of infection by amoeboid movement.

Monocytes are a kind of clean up crew.

- Monocytes phagocytize microorganisms, dead cells, cellular debris, and dead neutrophils.

Human Biology Fifth Edition Daniel D. Chiras — Section 7-4

Lymphocytes are involved in immune reaction to microorganisms.

- Lymphocytes are the second most numerous WBCs and play a vital role in immune protection.
- There are two types of lymphocytes: T lymphocytes and B lymphocytes.

Notes

Notes

Fifth Edition Human Biology Daniel D. Chiras — Section 7-4

Leukemia is a cancer of the white blood cells.

Infectious mononucleosis is a viral disease spread through saliva.

Fifth Edition Human Biology Daniel D. Chiras

Section 7-5 Blood Clotting

Platelets are a vital component of the blood-clotting mechanism.

- Clotting is a chain reaction stimulated by the release of a chemical called **thromboplastin**.
- Thrombin acts on another blood protein, **fibrinogen**.
- Fibrinogen is converted into fibrin.
- The fibrin web traps RBCs and platelets.
- Platelets captured by the fibrin web release additional thromboplastin.

Fifth Edition Human Biology Daniel D. Chiras — Section 7-5

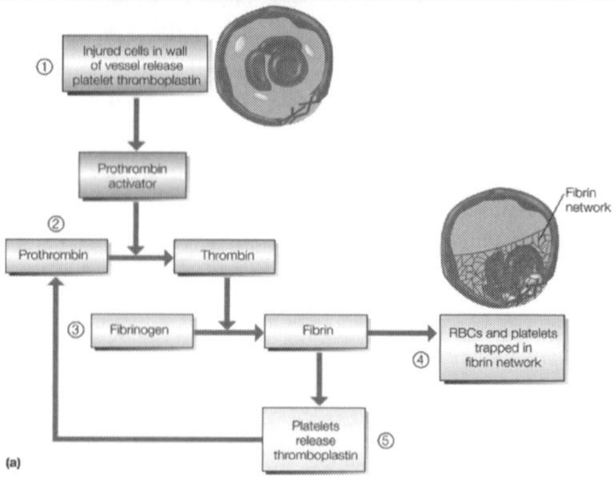

Figure 7-9 Blood Clotting Simplified

Fifth Edition Human Biology Daniel D. Chiras

Section 7-5

Hemophilia is a potentially life-threatening clotting disorder.

- In **hemophilia**, the liver fails to produce the necessary clotting factors.
- Hemophilia is caused by a hereditary genetic defect.

Fifth Edition Human Biology Daniel D. Chiras

Section 7-6 Blood Types

Blood types are determined by glycoproteins on RBC membranes.

- Four blood types exist: A, B, AB, and O.
- The letters refer to one type of glycoprotein present on the plasma membrane of RBCs.
- Serious problems arise when incompatible blood types are mixed.

The Rh factor must also be considered in blood transfusions.

- RBCs also contain other glycoproteins, notably the **Rh factor**, that can cause transfusion reactions.

Fifth Edition Human Biology Daniel D. Chiras

Section 7-7 Health and Homeostasis

The blood is vitally important to homeostasis.

- It transports materials, especially oxygen, to and from the cells.
- It protects against changes in pH.
- It transports excess heat to the body's surface.
- It plays a key role in the body's defense system.
- It seals injuries in blood vessels through clots.

Notes

Chapter 8: The Vital Exchange: Respiration

Notes

Fifth Edition Human Biology Daniel D. Chiras

Section 8-1 Structure of the Human Respiratory System

The respiratory system supplies oxygen to the body and gets rid of carbon dioxide.

The respiratory system consists of an air-conducting portion that delivers air to the lungs. and a gas-exchange portion.

Fifth Edition Human Biology Daniel D. Chiras — Section 8-1

The conducting portion of the respiratory system moves air in and out of the body and filters, warms, and moistens incoming air.

The alveoli are the site of gaseous exchange.

- The alveoli are tiny, thin-walled sacs formed by a single layer of flattened epithelial cells that facilitate diffusion.
- Surrounding the alveoli are capillary beds that pick up oxygen and expel carbon dioxide.

Notes

Fifth Edition Human Biology Daniel D. Chiras — Section 8-1

Figure 8-1 The Human Respiratory System

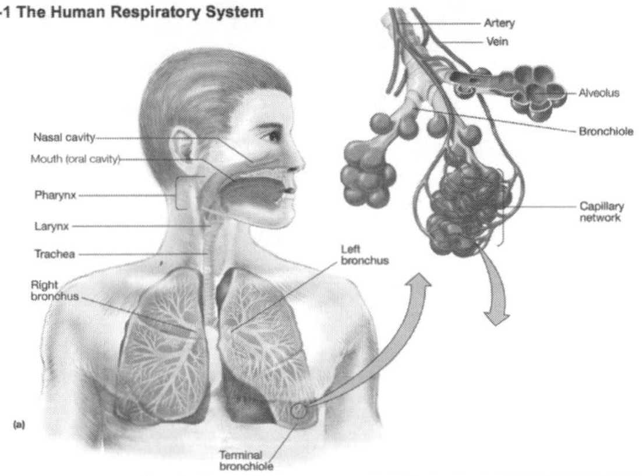

Fifth Edition Human Biology Daniel D. Chiras — Section 8-1

In humans, sound is produced by the vocal cords and is influenced by the tongue and oral cavity.

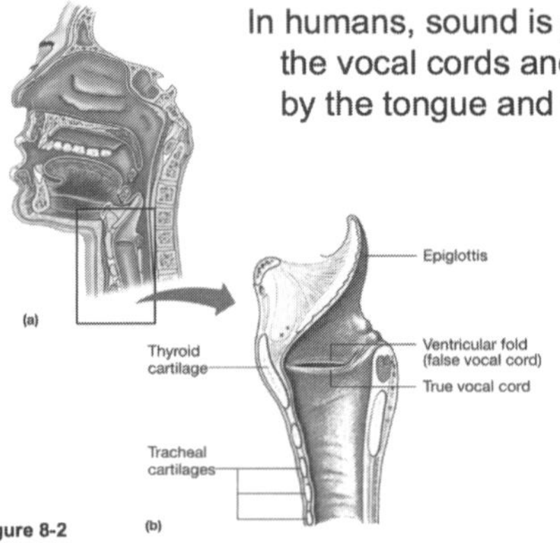

Figure 8-2

Fifth Edition Human Biology Daniel D. Chiras

Section 8-2 Breathing and the Control of Respiration

Air is moved in and out of the lungs as a result of changes in the pressure inside the chest cavity.

The health of a person's lungs can be assessed by measuring air flow in and out of them under various conditions.

Fifth Edition Human Biology Daniel D. Chiras — Section 8-2

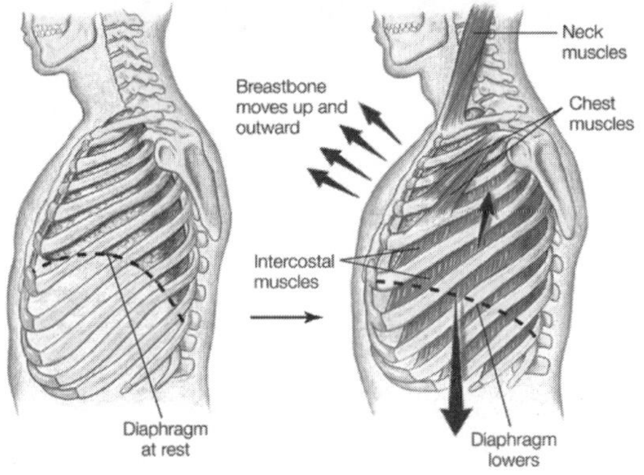

Figure 8-9 The Bellows Effect

Fifth Edition Human Biology Daniel D. Chiras — Section 8-2

Breathing is controlled principally by the breathing center in the brain.

- Nerve cells in the breathing center send impulses to the diaphragm and intercostal muscles, which contract.
- This increases the volume of the thoracic cavity, which draws air into the lungs.
- When the impulses stop, inspiration ends.
- Air is then expelled passively as the chest wall returns to the normal position, and the diaphragm rises.
- The breathing center is regulated by a negative feedback loop that it generates itself.

Fifth Edition Human Biology Daniel D. Chiras

Section 8-3 Gas Exchange in the Lungs

Oxygen and carbon dioxide diffuse rapidly across the alveolar and capillary walls.

- Oxygen diffuses into the blood plasma, then into the RBCs, where it binds to hemoglobin.
- Carbon dioxide enters the RBCs in the bloodstream, where it is converted into carbonic acid.
  - Carbonic acid dissociates, forming hydrogen and bicarbonate ions. The latter diffuse out of the RBCs.

Notes

Notes

Fifth Edition Human Biology Daniel D. Chiras

Section 8-3

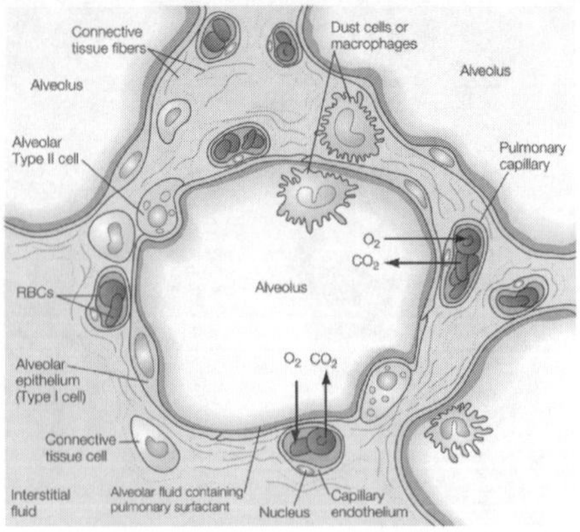

Figure 8-11 Close-up of the Alveolus

Fifth Edition Human Biology Daniel D. Chiras

Section 8-4 Diseases of the Respiratory System

Bacteria and viruses can infect many parts of the respiratory system.

Respiratory infections can spread to other organ systems.

The lungs are susceptible to airborne materials, such as asbestos.

Asthma is typically an allergic reaction to airborne irritants or other stimulus.

Chronic bronchitis is caused by an irritation of the lungs and bronchial passageways.

Fifth Edition Human Biology Daniel D. Chiras

Section 8-5 Health and Homeostasis

The respiratory system plays a vital role in homeostasis by regulating oxygen and carbon dioxide in the blood and body tissues.

Pathogenic organisms and chemical contaminants in the environment may harm the respiratory system.

Notes

Chapter 9

The Urinary System

Fifth Edition Human Biology Daniel D. Chiras

Section 9-1 Organs of Excretion

Cells of the body produce a number of wastes that must be excreted to maintain homeostasis.

Organisms must have a system for eliminating waste.

Fifth Edition Human Biology Daniel D. Chiras — Section 9-1

- Excretory organs remove waste and help regulate inorganic ions and water.
- The major metabolic wastes and substances include:
  - Ammonia
  - Urea
  - Uric acid
  - Bile pigments
  - Urochrome
  - Carbon dioxide
  - Water
  - Inorganic ions

Notes

Human Biology, Fifth Edition — Daniel D. Chiras

Section 9-2 The Urinary System

Humans have several organ systems that rid the body of wastes; one of the most important is the urinary system.

The urinary system consists of the kidneys, ureters, bladder, and urethra.

Human Biology, Fifth Edition — Daniel D. Chiras — Section 9-2

- Kidneys remove impurities from the blood and help regulate the water levels and ionic concentrations of the blood.
- Blood enters the kidneys in the renal arteries whose branches deliver it to millions of nephrons.
- The nephrons produce urine, which drains from the kidneys into the bladder via the ureters.
- Urine is voided through the urethra.

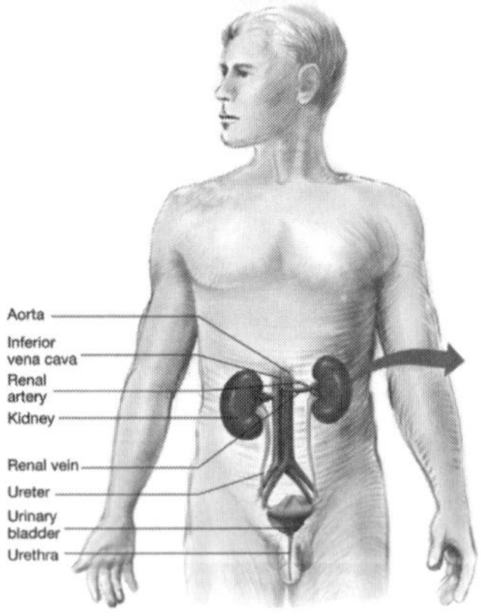

Figure 9-2

Human Biology, Fifth Edition — Daniel D. Chiras — Section 9-2

The human kidney consists of two zones, an outer cortex and inner medulla.

Nephrons consist of two parts, a glomerulus and a renal tubule.

- The renal tubule consists of four parts:
  - Bowman's capsule
  - The proximal convoluted tubule
  - The loop of Henle
  - The distal convoluted tubule

Fifth Edition Human Biology Daniel D. Chiras Section 9-2

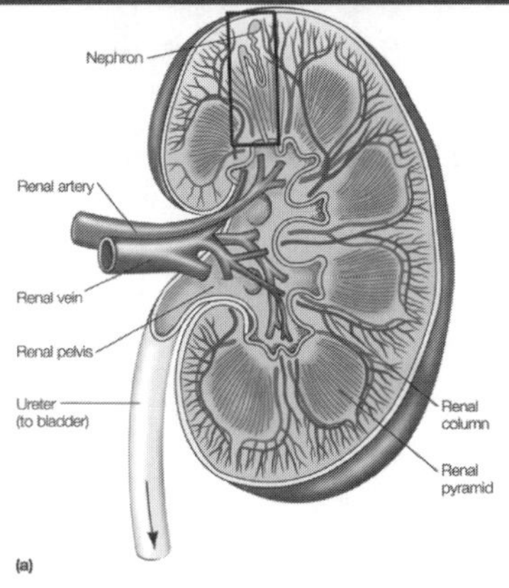

Figure 9–4 The Nephron -- Cross section of the kidney showing the location of the nephrons

Fifth Edition Human Biology Daniel D. Chiras Section 9-2

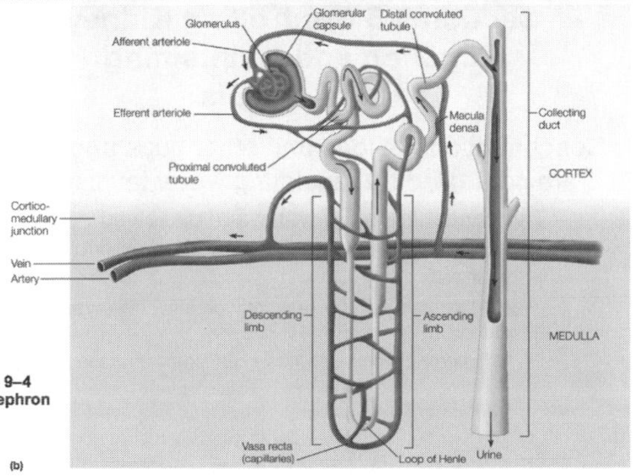

Figure 9–4 The Nephron

Fifth Edition Human Biology Daniel D. Chiras

Section 9-3 Function of the Urinary System

Blood filtration in nephrons involves three processes:

- Glomerular filtration is a physical filtration process
- Tubular reabsorption helps conserve valuable nutrients and ions.
- Tubular secretion is the transport of waste products from the peritubular capillaries into the renal tubule.

Notes

Notes

Fifth Edition Human Biology Daniel D. Chiras

Section 9-4 Controlling Urination

Urination is a reflex in very young children, but is controlled consciously by older children and adults.

Fifth Edition Human Biology Daniel D. Chiras

Section 9-5 Controlling Kidney Function and Maintaining Homeostasis

Concentrations of ions and other substances are controlled by regulating water levels.

- The concentration of water and dissolved substances in the blood is controlled by two hormones:
  - ADH increases water reabsorption and conserves body water.
  - Aldosterone stimulates water reabsorption in the kidney.

Fifth Edition Human Biology Daniel D. Chiras Section 9-5

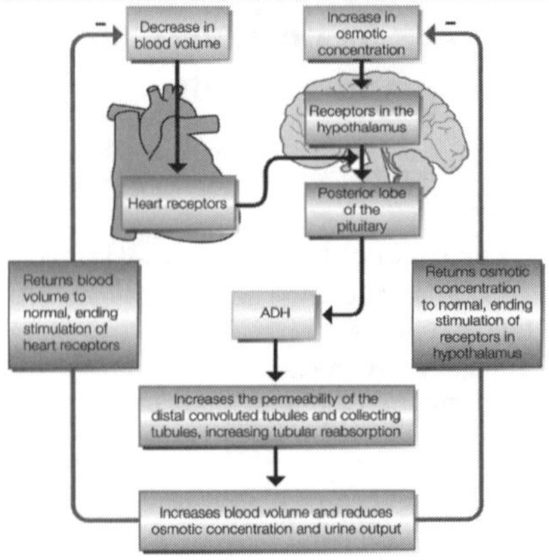

Figure 9-6 ADH Secretion

Fifth Edition Human Biology Daniel D. Chiras — Section 9-5

Water balance is also affected by several chemicals, like caffeine and alcohol (ethanol).

- Caffeine increases urine output without affecting ADH secretions.
- Ethanol inhibits the secretion of ADH by the pituitary.

Fifth Edition Human Biology Daniel D. Chiras

Section 9-6 Diseases of the Urinary System

Kidney stones can block the outflow of urine and cause severe kidney damage.

Renal failure may occur suddenly or gradually and can be treated by dialysis, a mechanical filtering of the blood.

Diabetes insipidus is a rare disease caused by a lack of ADH.

Fifth Edition Human Biology Daniel D. Chiras

Section 9-7 Health and Homeostasis

The kidneys help maintain the proper levels of nutrients and waste in the body. They also help eliminate toxins in the body and maintain blood Ph.

Kidneys can be damaged by severe blood loss, bacteria, and toxic chemicals.

Notes

Notes

Fifth Edition Human Biology Daniel D. Chiras

Section 10-1 An Overview of the Nervous System

The nervous system controls body functions, but also allows for higher functions such as thinking.

The nervous system consists of two parts, the central and peripheral nervous systems.

- The central nervous system (CNS) is made up of the brain and spinal cord.
- Receptors transmit sensory input to the CNS via sensory neurons.
- The CNS integrates all sensory input and generates appropriate responses.
- Motor output leaves the CNS in motor neurons.
- The peripheral nervous system (PNS) consists of the spinal and cranial nerves.

Fifth Edition Human Biology Daniel D. Chiras — Section 10-1

The peripheral consists of the somatic and the autonomic subdivisions.

- The autonomic division controls involuntary actions.
- The somatic division largely controls voluntary actions.

Notes

Fifth Edition Human Biology Daniel D. Chiras

Section 10-1

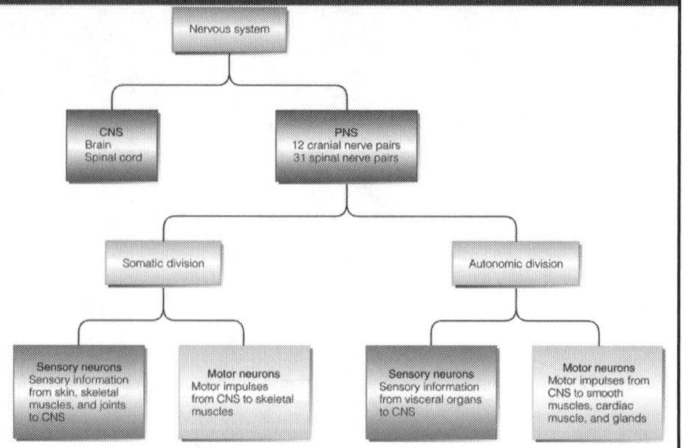

Figure 10-2 Subdivisions of the Nervous System

Fifth Edition Human Biology Daniel D. Chiras

Section 10-2 Structure and Function of the Neuron

The fundamental unit of the nervous system is the neuron.

This highly specialized cell generates and transmits bioelectric impulses.

Fifth Edition Human Biology Daniel D. Chiras

Section 10-2

The fundamental unit of the nervous system is the neuron.

This highly specialized cell generates and transmits bioelectric impulses.

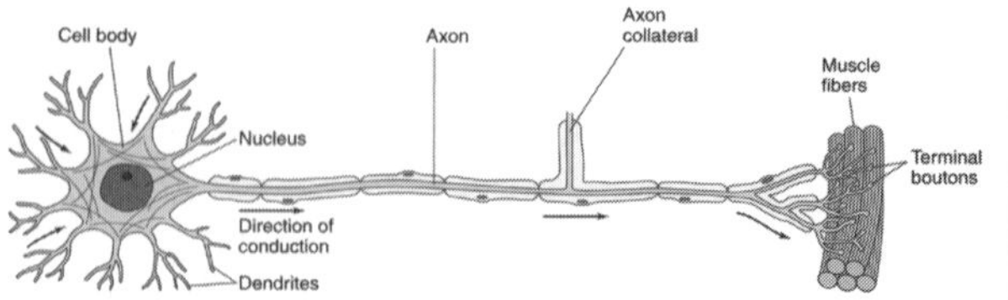

Figure 10-3b

Fifth Edition Human Biology Daniel D. Chiras — Section 10-2

All neurons consist of a cell body and two types of processes.

- Dendrites, which conduct impulses to the cell body
- Axons, which conduct impulses away from the cell body.
- Axons in the CNS and PNS contain an insulating layer, the myelin sheath, which greatly increases the rate of transmission of nerve impulses.

Fifth Edition Human Biology Daniel D. Chiras — Section 10-2

- The terminal ends of axons branch, forming fibers that end in small knobs called terminal boutons.
- Neurons lose the ability to divide.
- Neurons that die cannot be replaced by existing cells.
- Neurons have a high metabolic demand, making them highly susceptible to loss of oxygen and glucose.

Fifth Edition Human Biology Daniel D. Chiras — Section 10-2

Nerve impulses result from the flow of ions across their plasma membranes.

- The electrical potential across the membrane is known as the membrane potential or resting potential.
- When a nerve cell is stimulated, its plasma membrane increases its permeability to sodium ions.
- Sodium ions rush in, causing depolarization down the membrane.
- Depolarization is followed by repolarization.
- The depolarization and repolarization of the neuron's plasma membrane constitute a bioelectric impulse or action potential.

Notes

Notes

Fifth Edition Human Biology Daniel D. Chiras — Section 10-2

Nerve impulses travel from one neuron to another across synapses.

- When an impulse reaches the terminal bouton, it stimulates the release of neurotransmitters.
- Neurotransmitters may excite or inhibit the postsynaptic membrane.

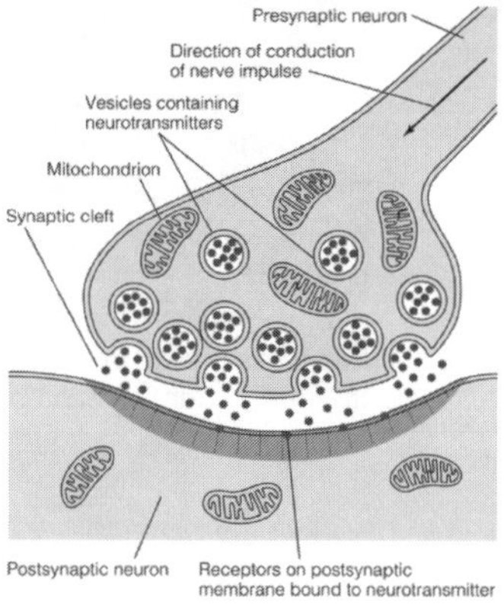

Figure 10-6b

Fifth Edition Human Biology Daniel D. Chiras

Section 10-3 The Spinal Cord and Nerves

Nerve cells can be grouped into three functional categories:

- Sensory neurons carry sensory impulses
- Motor neurons carry motor impulses
- Interneurons connect sensory and motor neurons in the spinal cord.

Fifth Edition Human Biology Daniel D. Chiras — Section 10-3

The spinal cord transmits information to and from the brain and houses many reflexes.

The nerves attached to the spinal cord carry motor and sensory impulses.

- Two types of nerves emanate from the CNS: spinal and cranial.
- Spinal nerves arise from the spinal cord and may be sensory, motor, or mixed.
- Cranial nerves attach to the brain and supply the structures of the head and several key body parts.

Fifth Edition Human Biology Daniel D. Chiras Section 10-3

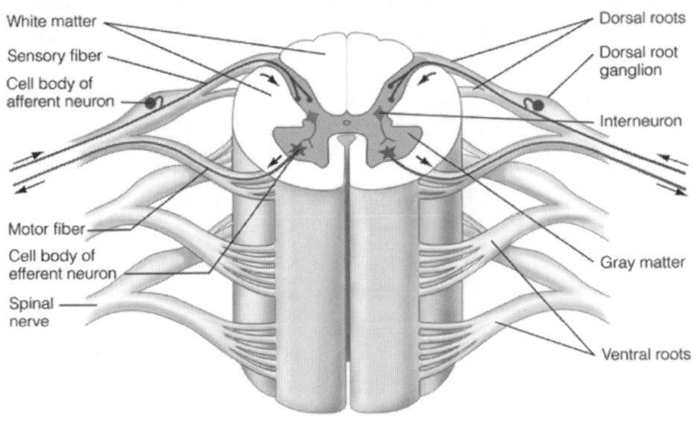

Figure 10-8 The Spinal Cord and Dorsal Root Ganglia

Fifth Edition Human Biology Daniel D. Chiras Section 10-3

Injury to the spinal cord can cause permanent damage.

- The severity of the injury depends on its location and the extent of the damage.

Fifth Edition Human Biology Daniel D. Chiras

Section 10-4 The Brain

The cerebral hemispheres function in integration, sensory reception, and motor action.

- The cerebrum with its two cerebral hemispheres is the largest part of the brain.
- The outer layer of each hemisphere is the cortex.
- The cerebral cortex consists of many discrete functional regions including motor, sensory, and association areas.
- The primary motor cortex controls voluntary movement.
- The primary sensory cortex receives sensory information from the body.
- The association cortex is the site of integration and complex intellectual activities.

Notes

Notes

Fifth Edition Human Biology Daniel D. Chiras — Section 10-4

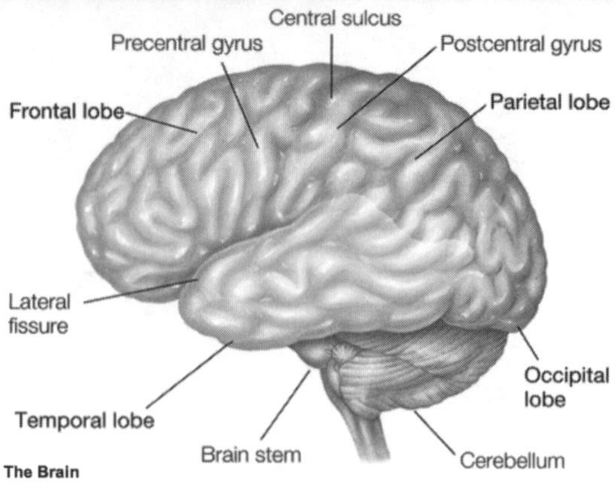

Figure 10-11 The Brain

Fifth Edition Human Biology Daniel D. Chiras — Section 10-4

Unconscious functions are housed in the cerebellum, hypothalamus, and brain stem.

- The cerebellum controls muscle synergy and helps maintain posture.
- The hypothalamus controls many autonomic functions involved in homeostasis.
- The limbic system is the site of instinctive behavior and emotion.
- The brain stem controls basic body functions.

Fifth Edition Human Biology Daniel D. Chiras — Section 10-4

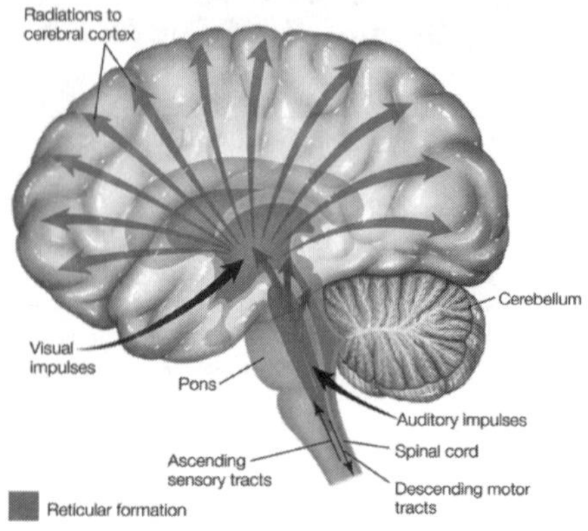

Figure 10-16 The Reticular Activating System

Fifth Edition Human Biology Daniel D. Chiras — Section 10-4

Cerebrospinal fluid cushions the CNS.

Electrical activity in the brain varies depending on activity level or level of sleep.

Headaches have many causes but are rarely the result of life-threatening anomalies.

Fifth Edition Human Biology Daniel D. Chiras

Section 10-5 The Autonomic Nervous System

The ANS controls many body functions not under conscious control.

The autonomic nervous system (ANS) is a division of the peripheral nervous system.

The ANS helps control internal organs through controls of smooth muscle, cardiac muscle, and glands.

The ANS has two subdivisions: the sympathetic and the parasympathetic.

- The sympathetic is responsible for the fight-or-fight response.
- The parasympathetic is responsible for internal responses associated with the relaxed state.

Fifth Edition Human Biology Daniel D. Chiras

Section 10-6 Learning and Memory

Learning is the acquisition of new information and skills; memory is the storage and recall of the at information depends on one's ability to store information in the brain.

Short-term memory retains information for periods of seconds to hours.

Long-term memory holds information for periods of days to years.

Notes

Notes

Human Biology Fifth Edition Daniel D. Chiras — Section 10-6

Memory is stored in multiple regions of the brain.

Short-term and long-term memory appear to involve structural and functional changes of the neurons.

New evidence suggests that brain function can be retained in old age.

Human Biology Fifth Edition Daniel D. Chiras

Section 10-7 Diseases of the Brain

Loss of memory and confusion in older individuals is often the result of underlying disease such as Alzheimer's disease.

Parkinson's disease is a neurological disorder caused by a lack of dopamine in certain regions of the brain.

Multiple sclerosis is caused by a destruction of the myelin sheaths of nerve cells in the CNS.

Brain tumors are either benign or malignant and may have many causes.

Human Biology Fifth Edition Daniel D. Chiras

Section 10-8 Health and Homeostasis

Uncontrolled blood glucose levels may cause changes in the brain and behavior.

Notes

Chapter 11

The Senses

Fifth Edition Human Biology Daniel D. Chiras

Section 11-1 The General and Special Senses

The body contains two types of senses:
general
and special.

Receptors for the senses are grouped into five functional categories:

- Mechanoreceptors
- Chemoreceptors
- Thermoreceptors
- Photoreceptors
- Nociceptors

Fifth Edition Human Biology Daniel D. Chiras

Section 11-2 The General Senses

Receptors for the general body senses generally fit into two groups:

- naked nerve endings
- encapsulated receptors.

Naked nerve endings in body tissues detect

- Pain
- Temperature
- Light touch

Notes

Fifth Edition Human Biology Daniel D. Chiras — Section 11-2

Encapsulated receptors consist of naked nerve endings surrounded by one or more layers of cells.

- Encapsulated receptors detect pressure, light touch, and muscle extension.

Figure 11-1

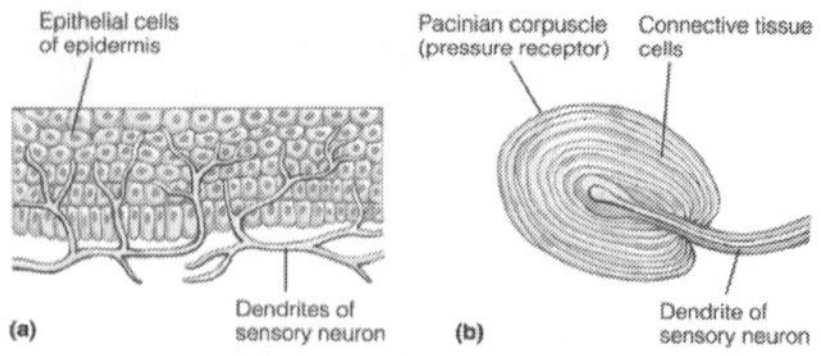

Many receptors stop generating impulses after exposure to a stimulus for some length of time.

Receptors play an important role in homeostasis.

Fifth Edition Human Biology Daniel D. Chiras

Section 11-3 Taste and Smell

The special senses include

- Taste
- Smell
- Vision
- Hearing
- Balance

Fifth Edition Human Biology Daniel D. Chiras — Section 11-3

Taste buds respond to chemicals dissolved in food.

- Located principally on the upper surface of the tongue.
- Food molecules dissolve in the saliva and bind to the membranes of the microvilli of the receptor cells.
- Taste buds respond to five flavors: salty, bitter, sweet, sour, and umami.
  - Taste buds are preferentially responsive to one flavor.

Fifth Edition
Human Biology Daniel D. Chiras Section 11-3

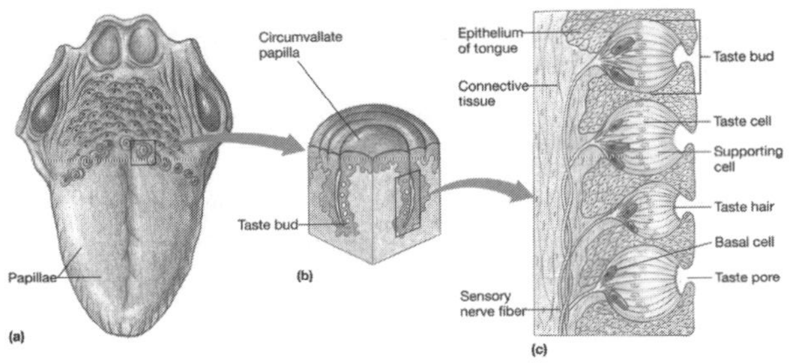

Figure 11-4 Taste Receptors

Fifth Edition
Human Biology Daniel D. Chiras Section 11-3

The olfactory epithelium is a patch of receptor cells that detects odors.

- Located in the roof of the nasal cavities.
- The receptor cells respond to thousands of different molecules.
- The molecules bind to membrane receptors on the olfactory hairs, stimulating nerve impulses that are transmitted to the brain via the olfactory nerve.

Smell influences our sense of taste, and vice versa.

Fifth Edition
Human Biology Daniel D. Chiras Section 11-3

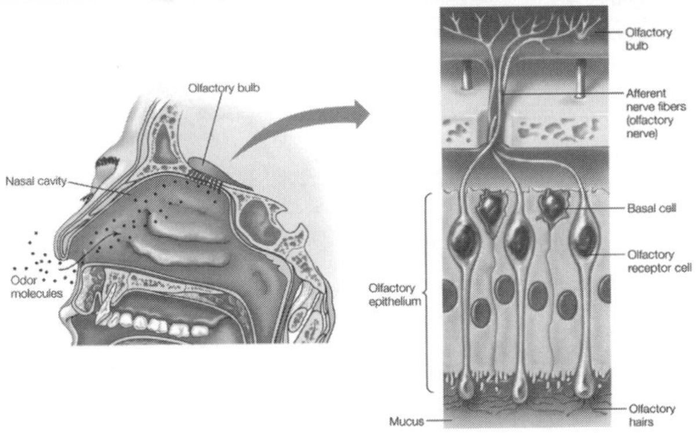

Figure 11-5 The Olfactory Epithelium

Notes

Notes

Fifth Edition Human Biology Daniel D. Chiras

Section 11-4 The Visual Sense: The Eye

The human eye consists of three distinct layers.

- The outermost layer consists of the sclera and the cornea.
- The middle layer consists of the choroid, the ciliary body, and the iris.
- The innermost layer is the retina, the light-sensitive layer,
  - The retina contains two types of photoreceptors: rods and cones.
    - Rods function in dim light and provide black-and-white vision
    - Cones operate in bright light and provide color vision.
    - Cones are also responsible for visual acuity.

Fifth Edition Human Biology Daniel D. Chiras Section 11-4

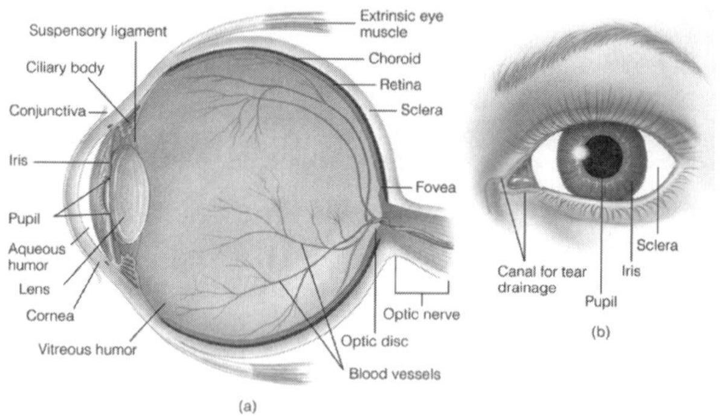

Figure 11-6 Anatomy of the Human Eye

Fifth Edition Human Biology Daniel D. Chiras Section 11-4

Figure 11-7 The Retina

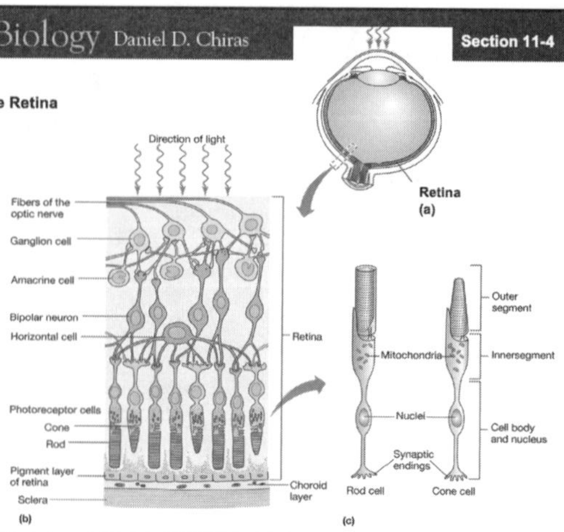

Human Biology Fifth Edition Daniel D. Chiras — Section 11-4

- The lens and cornea focus light on the retina.
- Alterations in the shape of the lens and eyeball cause the most common visual problems.
- Overlapping visual fields give us depth perception.

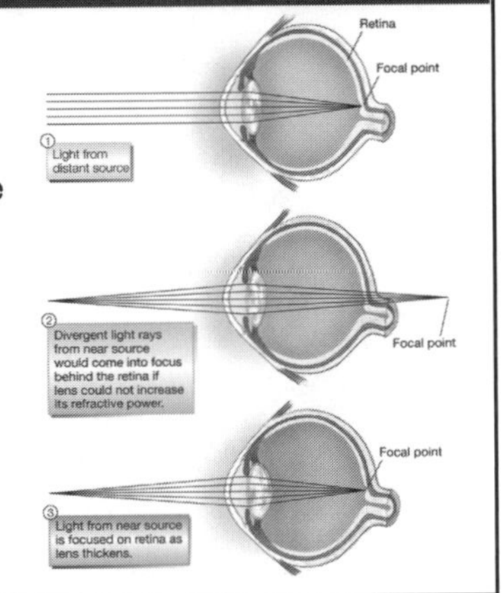

Figure 11–8

Human Biology Fifth Edition Daniel D. Chiras

Section 11-5 Hearing and Balance: The Cochlea and Middle Ear

The ear consists of three anatomically separate portions: the outer, middle, and inner ears.

- The outer ear consists of the auricle and external auditory canal, both of which direct sound to the eardrum.
- The middle ear consists of the eardrum and three small bones, the ossicles, which transmit vibrations to the inner ear.
- The auditory tube equilibrates the pressure inside the middle ear cavity.
- The inner ear contains the cochlea where the receptors for sound are located.

Human Biology Fifth Edition Daniel D. Chiras — Section 11-5

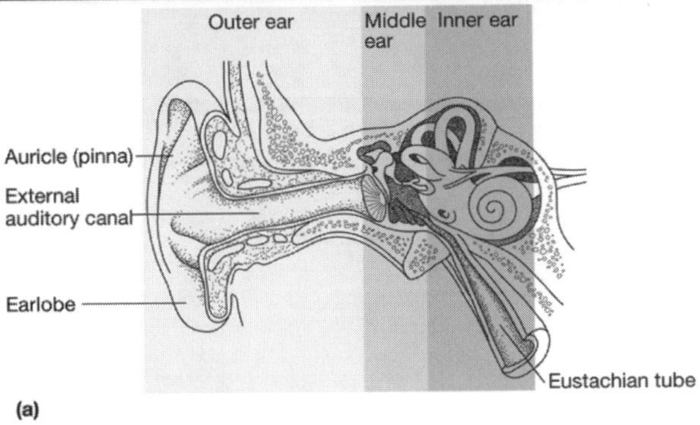

Figure 11-13 The Structurs of the Ear

Notes

Notes

Fifth Edition Human Biology Daniel D. Chiras

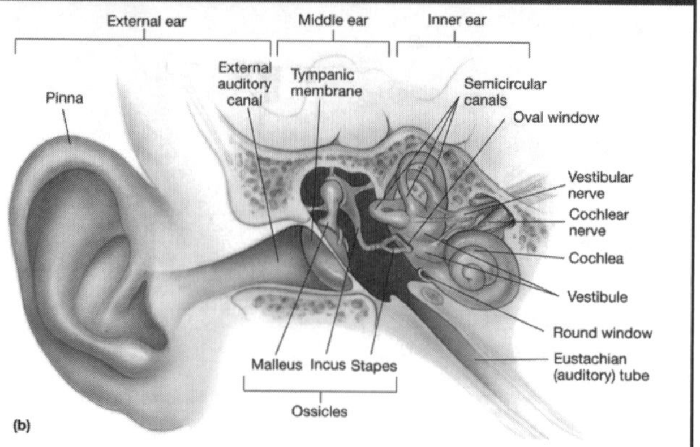

Figure 11-13 The Structurs of the Ear

Fifth Edition Human Biology Daniel D. Chiras — **Section 11-5**

The Eustachian tube helps to equalize pressure in the middle ear.

Hearing requires the participation of several structures.

- The cochlea is a spiral-shaped, bony structure that contains three fluid-filled canals. Separating the middle canal from the lower one is the flexible basilar membrane that supports the organ of Corti.
  - Hair cells in the organ of Corti are embedded in the relatively rigid tectorial membrane.
- Sound waves create vibrations in the eardrum and ossicles, which are transmitted to fluid in the cochlea.

Fifth Edition Human Biology Daniel D. Chiras — **Section 11-5**

- Pressure waves in the cochlea cause the basilar membrane to vibrate, which stimulates the hair cells.
- Pressure waves resulting from a sound cause one part of the membrane to vibrate maximally.
- The hair cells stimulated in that region send signals to the brain, which it interprets as a specific frequency.

Fifth Edition Human Biology Daniel D. Chiras — Section 11-5

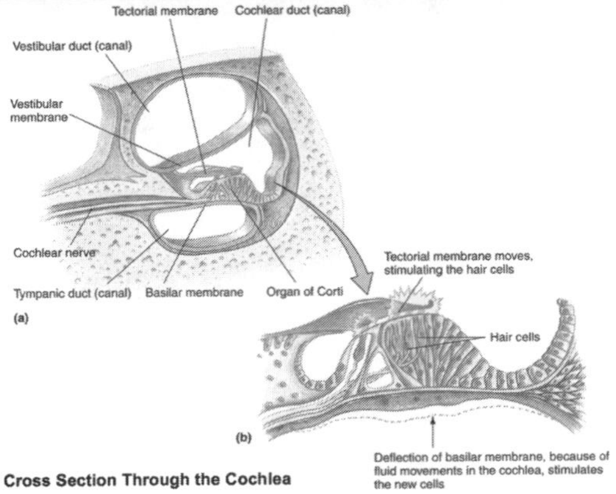

Figure 11-14 Cross Section Through the Cochlea

Fifth Edition Human Biology Daniel D. Chiras — Section 11-5

The vestibular apparatus houses receptors that detect body position and movement.

- The semicircular canals, utricle, and saccule.
- Fluid movement inside the semicircular canals deflects the cupula lying over the receptor cells, alerting the brain to head movements.
- The utricle and saccule contain receptors that respond to linear acceleration and tilting of the head.

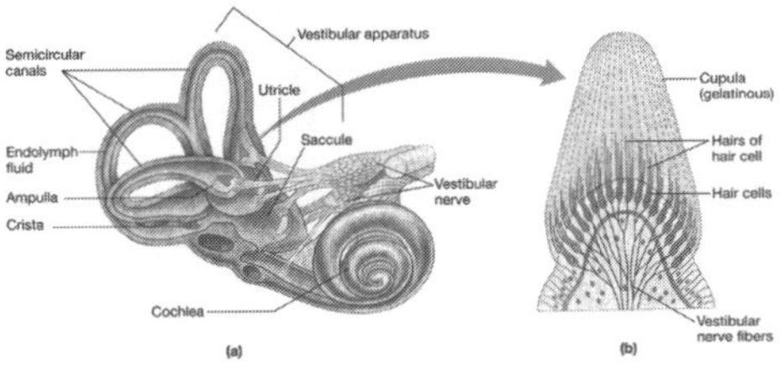

Figure 11-16

Fifth Edition Human Biology Daniel D. Chiras

Section 11-6 Health and Homeostasis

Noise pollution in our environment affects homeostasis by disturbing sleep and raising stress levels.

Notes

Notes

Chapter 12

The Skeleton and Muscles

Human Biology Fifth Edition Daniel D. Chiras

Section 12-1 Structure and Function of the Human Skeleton

Bones serve many functions and play an important role in homeostasis. Bones

- Provide internal support.
- Allow for movement.
- Help protect internal body parts.
- Produce blood cells and platelets.
- Store fat.
- Help regulate blood calcium levels.

Human Biology Fifth Edition Daniel D. Chiras — Section 12-1

The human skeleton consists of two parts:

- The axial skeleton
- The appendicular skeleton

Notes

Fifth Edition Human Biology Daniel D. Chiras — Section 12-1

All bones have a hard, dense outer layer that surrounds a less compact central region.

- Inside the bone is the marrow cavity, filled with either fat cells (yellow marrow) or blood cells and blood-producing cells (red marrow) or with combinations of the two.

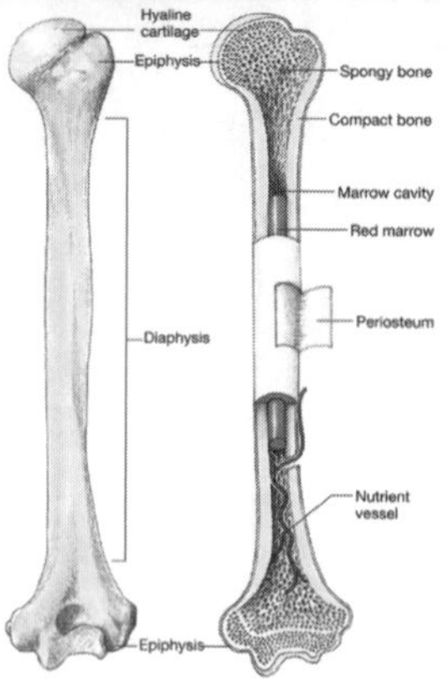

Figure 12-3

Fifth Edition Human Biology Daniel D. Chiras — Section 12-1

Bones are constantly remodeled in adults to meet changing stresses placed on them.

- Two cells are responsible for bone remodeling:
  - Osteoclasts—destroy bone
  - Osteoblasts—form bone

Bone is a homeostatic organ that helps maintain proper levels of calcium in the body.

Fifth Edition Human Biology Daniel D. Chiras — Section 12-1

The joints permit varying degrees of mobility.

- Immovable
- Slightly movable
- Freely movable

Most of the bones of the human skeleton start out as hyaline cartilage.

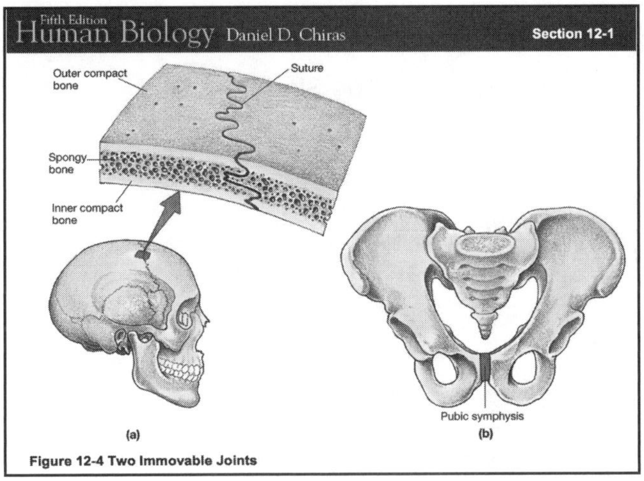

Figure 12-4 Two Immovable Joints

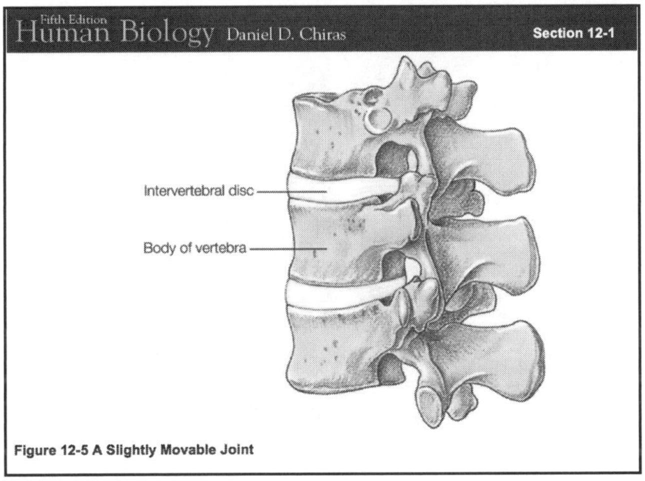

Figure 12-5 A Slightly Movable Joint

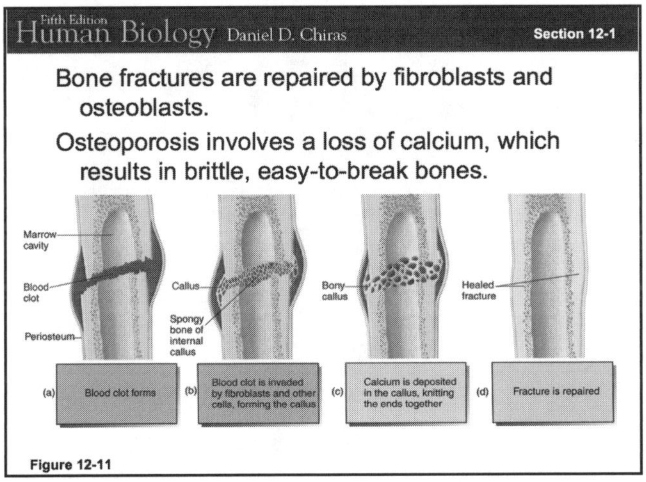

Figure 12-11

Notes

Notes

Fifth Edition Human Biology Daniel D. Chiras

Section 12-2 The Skeletal Muscles

Skeletal muscles are involved in

Body movements
Maintenance of posture
Production of body heat

Fifth Edition Human Biology Daniel D. Chiras

Section 12-2

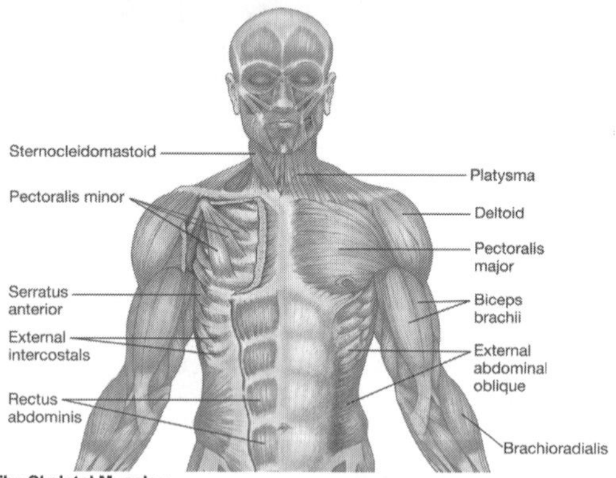

Figure 12-13 The Skeletal Muscles

Fifth Edition Human Biology Daniel D. Chiras

Section 12-2

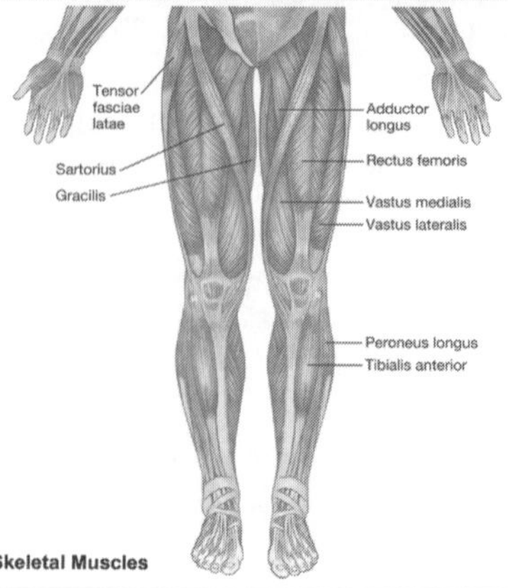

Figure 12-13 The Skeletal Muscles

Fifth Edition Human Biology Daniel D. Chiras — Section 12-2

Figure 12-13 The Skeletal Muscles

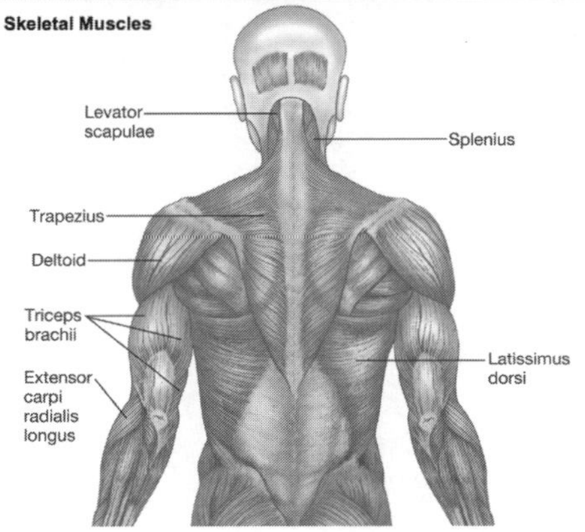

Fifth Edition Human Biology Daniel D. Chiras — Section 12-2

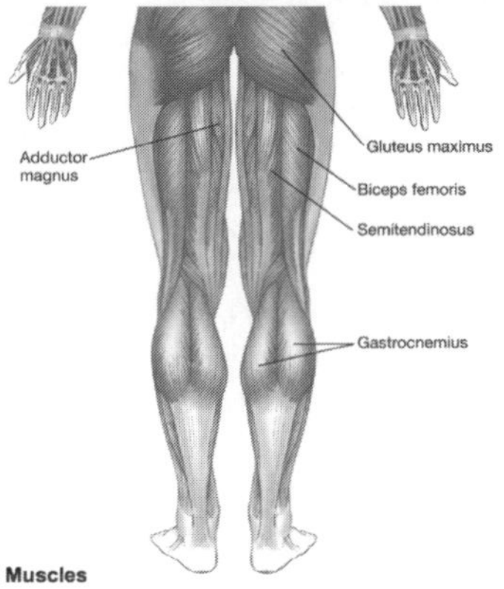

Figure 12-13 The Skeletal Muscles

Fifth Edition Human Biology Daniel D. Chiras — Section 12-2

Skeletal muscle cells are known as muscle fibers and are both excitable and contractile.

Muscle fibers contain many small bundles of contractile filaments known as myofibrils.

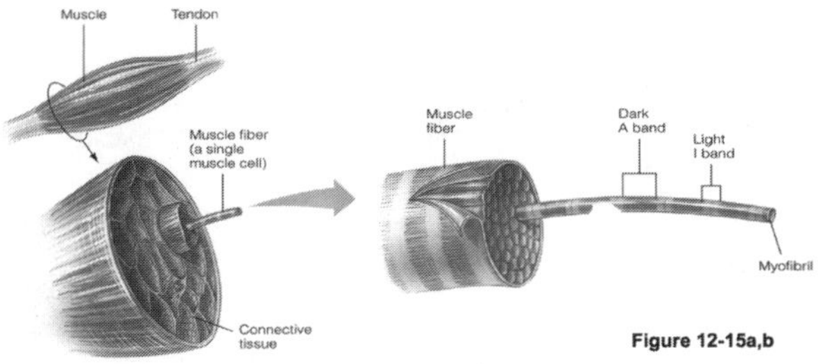

Figure 12-15a,b

Notes

Notes

Fifth Edition Human Biology Daniel D. Chiras — Section 12-2

Figure 12-15c-e

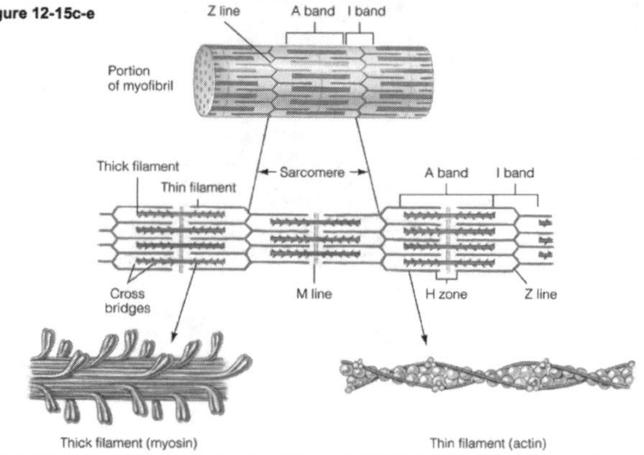

Fifth Edition Human Biology Daniel D. Chiras — Section 12-2

During muscle contraction, the actin filaments slide inward, causing the sarcomeres to shorten.

- The energy for muscle contraction comes from ATP.
- ATP is replenished by creatine phosphate, glycolysis, and cellular respiration.

Fifth Edition Human Biology Daniel D. Chiras — Section 12-2

Individual skeletal muscle fibers contract after being stimulated by an action potential.

- The strength of muscle contraction can be increased by stimulating (recruiting) additional muscle fibers to contract.
- Additional tension may be created in a muscle fiber if a nerve impulse arrives while the muscle fiber is still contracted.

Fifth Edition Human Biology Daniel D. Chiras — Section 12-2

Muscle tone results from the contraction of a small number of muscle fibers that keep muscles slightly tense.

Two types of muscle fibers are found in skeletal muscle, slow- and fast-twitch.

- Fast-twitch fibers are found in greatest number in muscles that perform rapid movement.
- Slow-twitch fibers are found in muscles that perform slower motions or are involved in maintaining posture.

Fifth Edition Human Biology Daniel D. Chiras — Section 12-2

Exercise builds muscles and increases endurance.

- An increase in muscle mass results from an increase in the amount of contractile protein in muscle fibers.
- Endurance is a function of at least three factors: the condition of the heart, the condition of the muscles of inspiration, and the blood volume.
    - Improvement in all three factors increases the efficiency of oxygen delivery to muscles.

Fifth Edition Human Biology Daniel D. Chiras

Section 12-3 Health and Homeostasis

Intentional misuse of steroids can cause physical, psychiatric, and behavioral problems.

Notes

Notes

Fifth Edition Human Biology Daniel D. Chiras

Section 13-1 Principles of Endocrinology

The endocrine system produces hormones that are transported in the blood to distant sites where they influence many functions.

Hormones affect five vital aspects of our lives:

Homeostasis
Growth and development
Reproduction
Energy production, storage, and use
Behavior

Fifth Edition Human Biology Daniel D. Chiras — Section 13-1

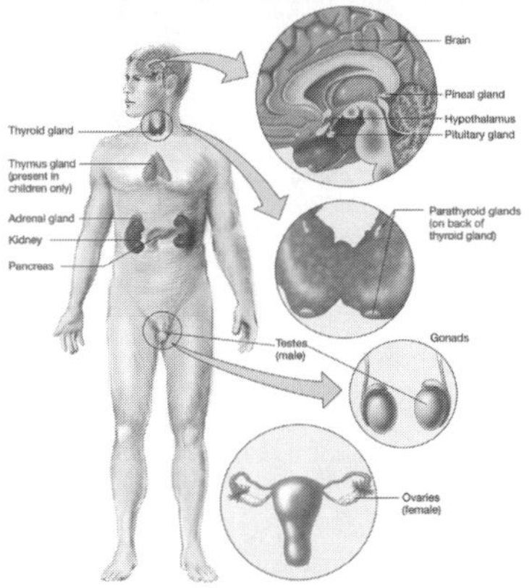

Figure 13-1 The Human Endocrine System

Notes

Fifth Edition Human Biology Daniel D. Chiras — Section 13-1

Target cells contain receptors for specific hormones.

Hormones stimulate the synthesis and release of other hormones or activate cellular processes.

- Three types of hormones are produced in the body:
  - Steroids
  - Proteins and polypeptides
  - Amines

Fifth Edition Human Biology Daniel D. Chiras — Section 13-1

Hormone secretion is often controlled by negative feedback mechanisms.

Most hormones undergo periodic fluctuations in their release.

The chemical nature of a hormone determines how it is transported in the blood and how it acts on cells.

Fifth Edition Human Biology Daniel D. Chiras — Section 13-1

The endocrine system and the nervous system are both control systems, but they operate in markedly different ways.

- The nervous system elicits rapid, generally short-lived responses.
- The endocrine system elicits slower, longer-lasting responses.

Fifth Edition Human Biology Daniel D. Chiras

Section 13-2 The Pituitary and Hypothalamus

The pituitary is a pea-sized gland suspended from the hypothalamus by a thin stalk.

It consists of two parts:

The anterior pituitary
The posterior pituitary

Fifth Edition Human Biology Daniel D. Chiras — Section 13-2

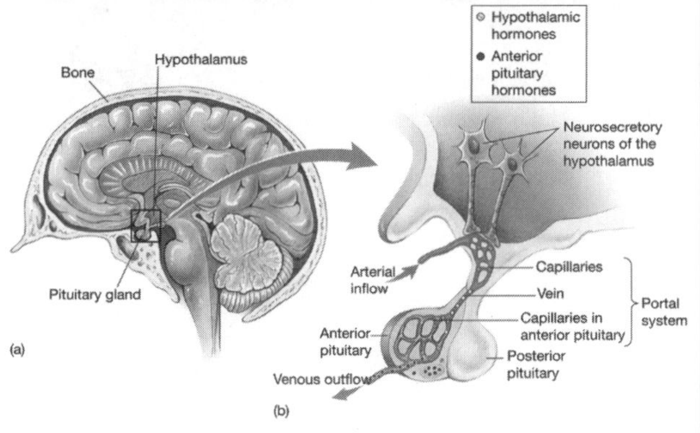

Figure 13-4a, b The Pituitary Gland

Fifth Edition Human Biology Daniel D. Chiras — Section 13-2

- The anterior pituitary secretes seven hormones with widely different functions.
- Growth hormones stimulate cell growth, primarily targeting muscle and bone.
- Thyroid-stimulating hormone stimulates the thyroid gland to produce thyroxine.
- ACTH stimulates the release of hormones from the adrenal cortex.
- The gonadotropins are hormones that stimulate the ovaries and testes.
- Prolactin stimulates milk production in the breasts of women.

Notes

Notes

Fifth Edition Human Biology Daniel D. Chiras | Section 13-2

The posterior pituitary secretes two hormones.

- Antidiuretic hormone increases the water absorption in the kidney.

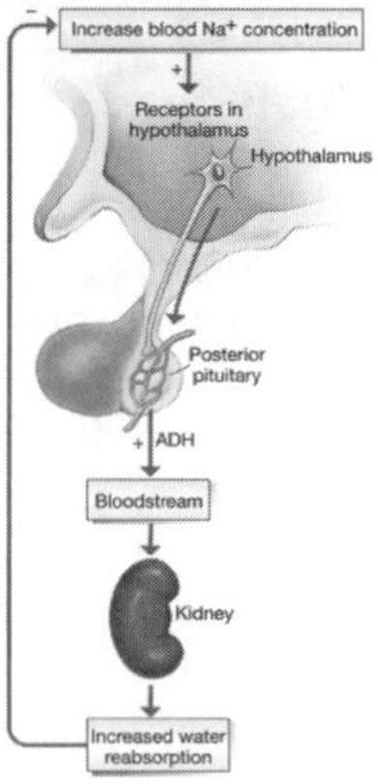

Figure 13-12

Fifth Edition Human Biology Daniel D. Chiras | Section 13-2

Oxytocin facilitates birth and stimulates milk let-down.

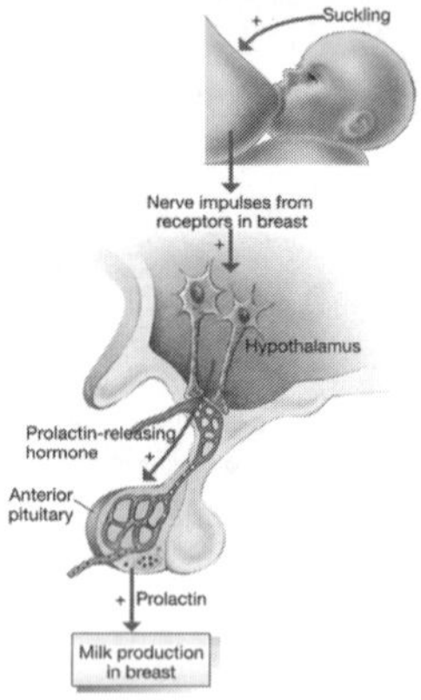

Figure 13-10

Fifth Edition Human Biology Daniel D. Chiras

Section 13-3 The Thyroid Gland

The thyroid gland located in the neck on either side of the trachea near its junction with the larynx.

The thyroid produces three hormones:

Thyroxine (T4)
Triiodothyronine
Calcitonin

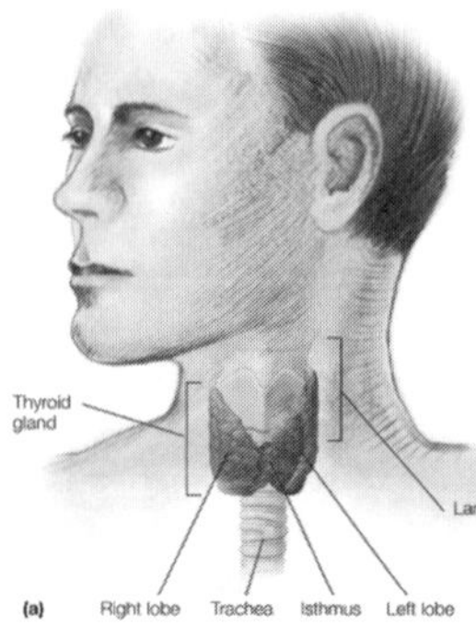

Figure 13-13a

Fifth Edition Human Biology Daniel D. Chiras | **Section 13-3**

Thyroxine and triiodothyronine accelerate the breakdown of glucose and stimulate growth and development.

Calcitonin decreases blood levels of calcium.

Fifth Edition Human Biology Daniel D. Chiras

Section 13-4 The Parathyroid Glands

The parathyroid glands produce the parathyroid hormone (PTH).

- PTH increases blood calcium levels.

Fifth Edition Human Biology Daniel D. Chiras

Section 13-5 The Pancreas

The pancreas produces two hormones, insulin and glucagon, from the islets of Langerhans.

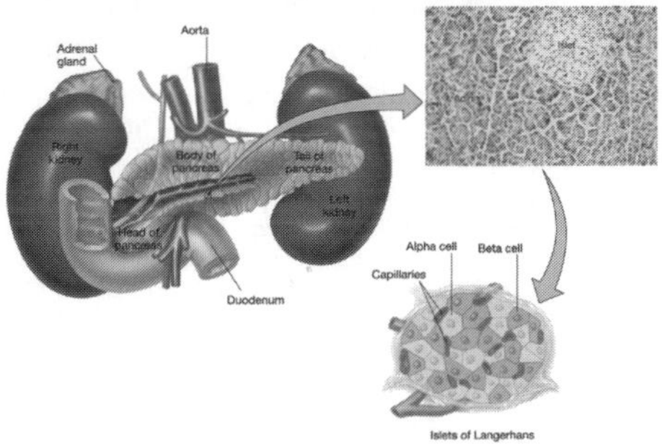

Figure 13-15

Notes

Notes

Fifth Edition **Human Biology** Daniel D. Chiras — **Section 13-5**

Insulin is a glucose-storage hormone and is produced by the beta cells. Insulin causes blood sugar levels to decline rapidly after a meal.

Glucagon increases blood levels of glucose, thus opposing the actions of insulin.

Fifth Edition **Human Biology** Daniel D. Chiras — **Section 13-5**

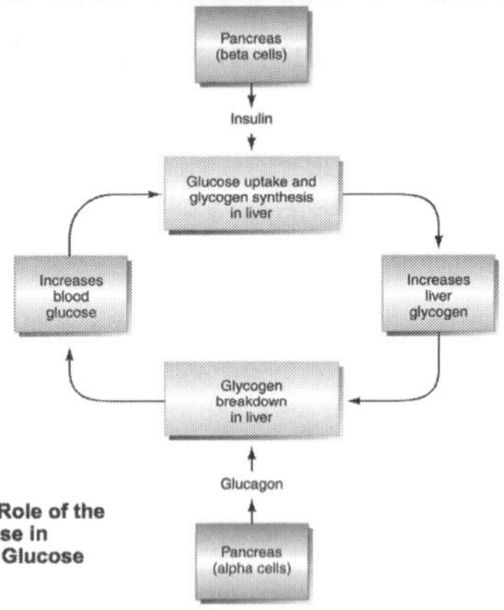

Figure 13-16 The Role of the Liver and Pancreas in Controlling Blood Glucose Levels

Fifth Edition **Human Biology** Daniel D. Chiras — **Section 13-5**

Diabetes mellitus is a disease resulting from an insulin deficiency or a decrease in tissue sensitivity to insulin.

- Diabetes has two principal forms:
  - Type I, or early-onset diabetes, may be caused by an autoimmune reaction that destroys the beta cells of the pancreas. It can be treated by insulin injections.
  - Type II, or late-onset diabetes, results from a reduction in the number of insulin receptors on target cells. It is caused by obesity and genetic factors and can often be treated successfully by dietary management.

Notes

Fifth Edition Human Biology Daniel D. Chiras

Section 13-6 The Adrenal Glands

The adrenal glands lie atop the kidneys and consist of an inner and outer regions, both of which produce hormones:

The adrenal medulla
The adrenal cortex

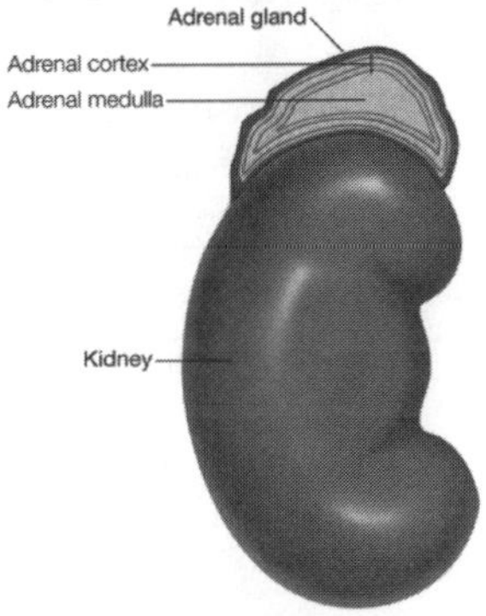

Figure 13-18

Fifth Edition Human Biology Daniel D. Chiras — Section 13-6

The adrenal medulla produces stress hormones.

- Adrenalin and noradrenalin stimulate heart rate and breathing, elevate blood glucose levels, constrict blood vessels in the intestine, and dilate blood vessels in the muscles.

Fifth Edition Human Biology Daniel D. Chiras — Section 13-6

The adrenal cortex produces three types of hormones with markedly different functions.

- Glucocorticoids, mineralocorticoids, and sex steroids.
  - Glucocorticoids affect carbohydrate metabolism and tend to raise blood glucose levels. The principal glucocorticoid is cortisol.
  - The chief mineralocorticoid is aldosterone. It acts on the kidneys, sweat glands, and salivary glands causing sodium and water retention and potassium excretion.

Notes

Fifth Edition Human Biology Daniel D. Chiras

Section 13-7 Health and Homeostasis

The endocrine system is sensitive to outside factors.

Stress can lead to an imbalance in the adrenal hormones.

Toxic pollutants can affect hormonal balance.

Notes

Human Biology Fifth Edition Daniel D. Chiras

Section 14-1 Viruses and Bacteria
An Introduction

Viruses are nonliving biological agents that invade cells.

Viruses consist of
- a nucleic acid core
- a coat of protein molecules called the capsid.

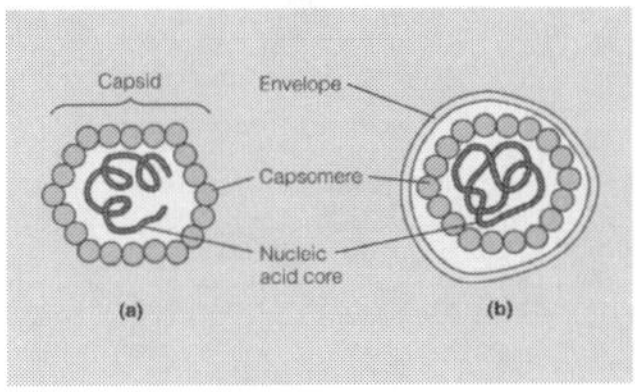

Figure 14-1ab

Human Biology Fifth Edition Daniel D. Chiras — Section 14-1

Many bacteria perform useful functions, while others cause serious diseases.

Bacteria are classified living organisms that reproduce on their own without taking over host cells.

Bacteria respond to antibiotics although many have become resistant to commonly used antibiotics; viruses do not, but new antiviral drugs may help flu sufferers.

Notes

Fifth Edition Human Biology Daniel D. Chiras

Section 14-2 The First and Second Lines of Defense

In humans, the first line of defense is a physical and chemical barrier.

Physical

- the skin
- epithelial linings of the respiratory, digestive, and urinary systems

Chemical

- Body secretions that impair bacterial growth

Fifth Edition Human Biology Daniel D. Chiras Section 14-2

The body's second line of defense combats infectious agents that penetrate the first line and consists of cellular and chemical responses.

Four nonspecific mechanisms include the imflammatory response, pyrogens, interferons, and complement.

Fifth Edition Human Biology Daniel D. Chiras Section 14-2

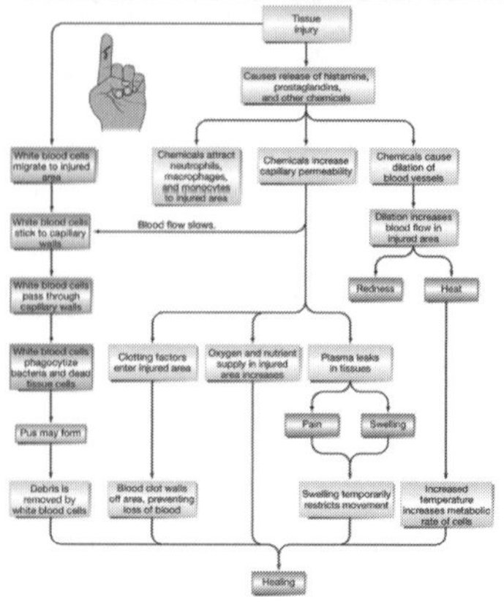

Figure 14-3 The Inflammatory Response

Fifth Edition Human Biology Daniel D. Chiras — Section 14-2

The inflammatory response involves a chemical and cellular responses.

– Macrophages, cells derived from the monocyte, are found in connective tissue beneath epithelia, where they phagocytize infectious agents, preventing their spread.

– Histamine, a chemical released by damaged tissue, stimulates arterioles in the infected tissue to dilate.

– The increase in blood flow raises the temperature of the wound.

– Heat stimulates macrophage metabolism, accelerating the rate of the destruction of infectious agents.

Fifth Edition Human Biology Daniel D. Chiras — Section 14-2

– Other chemicals increase the permeability of the capillaries, causing plasma to flow into the wound.

The increase in blood flow, the release of chemical attractants, and the flow of plasma into the wound constitute the inflammatory response.

Fifth Edition Human Biology Daniel D. Chiras — Section 14-2

The second line of defense consists of three additional chemicals, pyrogens, interferons, and complement.

– Pyrogens are molecules released primarily by macrophages exposed to bacteria, which raise body temperature and lower iron availability, thus decreasing bacterial replication.

– Interferons, a group of proteins released by cells infected by viruses, travel to other virus-infected cells, where they inhibit viral replication.

– The blood also contains the complement proteins, which circulate in an inactive state, becoming activated only when the body is invaded by bacteria.

Notes

Notes

Fifth Edition Human Biology Daniel D. Chiras

Section 14-2

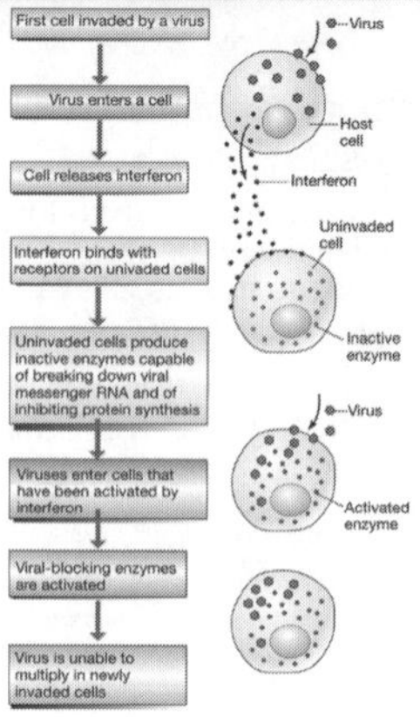

Figure 14-4
How Interferon Works

Fifth Edition Human Biology Daniel D. Chiras

Section 14-3 The Third Line of Defense: The Immune System

The cells of the immune system, lymphocytes, selectively target foreign substances and foreign organisms.

Lymphocytes circulate in the blood and lymph and take up residence in the lymphoid organs and lymphoid tissues.

Fifth Edition Human Biology Daniel D. Chiras

Section 14-3

Substances that trigger an immune response are large molecular weight molecules, particularly proteins and polysaccharides, called antigens.

- Antigens are antibody-generating substances.

Fifth Edition Human Biology Daniel D. Chiras — Section 14-3

Immature B and T cells are incapable of responding to antigens but soon gain this ability.

- B cells mature and differentiate in the bone marrow.
- T cells mature in the thymus.

Fifth Edition Human Biology Daniel D. Chiras — Section 14-3

B cells provide humoral immunity through the production of antibodies.

- Antibodies are proteins that bind to specific antigens.
- When T and B cells first encounter an antigen, the primary response, they react slowly.
- The rapidity of the secondary response is the result of the production of memory cells during the primary response.

Fifth Edition Human Biology Daniel D. Chiras — Section 14-3

Antibodies destroy antigens in one of four ways:

- Neutralization
- Agglutination
- Precipitation
- Activation of the complement system

Notes

Notes

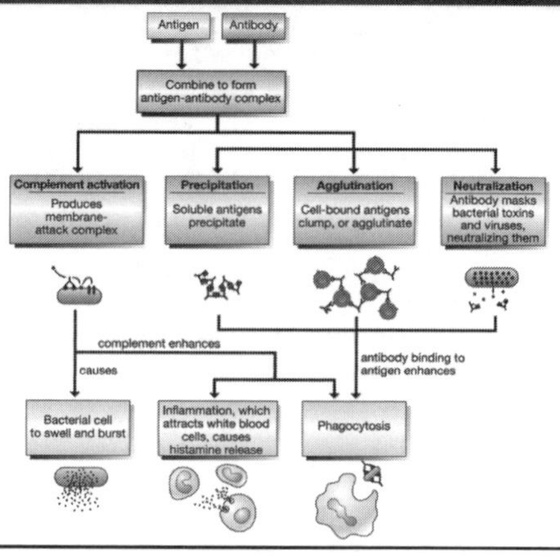

Figure 14-9 How Antibodies Work

Fifth Edition Human Biology Daniel D. Chiras Section 14-3

T cells attack foreign cells directly.

T cells differentiate into at least four cell types, each with a separate function in cell-mediated immunity.

- Cytotoxic T cells
- Memory T cells
- Helper T cells
- Suppressor T cells

Fifth Edition Human Biology Daniel D. Chiras Section 14-3

Macrophages play a very important role in humoral and cell-mediated immunity.

Two types of immunity are possible: active and passive.

- A vaccine containing a dead or weakened virus, bacterium, or bacterial toxin is injected into people to create active immunity.
- Passive immunity can be achieved by injecting antibodies into a patient or by the transfer of antibodies from a mother to her baby.

Fifth Edition
Human Biology Daniel D. Chiras

Section 14-3

The immune system may operate optimally when exposed to antigens early on.

Tissue transplantation often evokes cell-mediated immunity, which can be blocked by certain drugs.

Notes

Fifth Edition
Human Biology Daniel D. Chiras

Section 14-4 Diseases of the Immune System

The malfunctions of the immune system include allergies and autoimmune diseases.

Allergies are an immune system overreaction to environmental substances.

- Allergies are caused by IgE antibodies, produced by plasma cells.
- IgE antibodies bind to mast cells, which causes them to release histamine and other chemical substances that induce the symptoms of an allergy—production of mucus, sneezing, and itching.

Autoimmune diseases result from an immune attack on the body's own cells.

Fifth Edition
Human Biology Daniel D. Chiras

Section 14-4

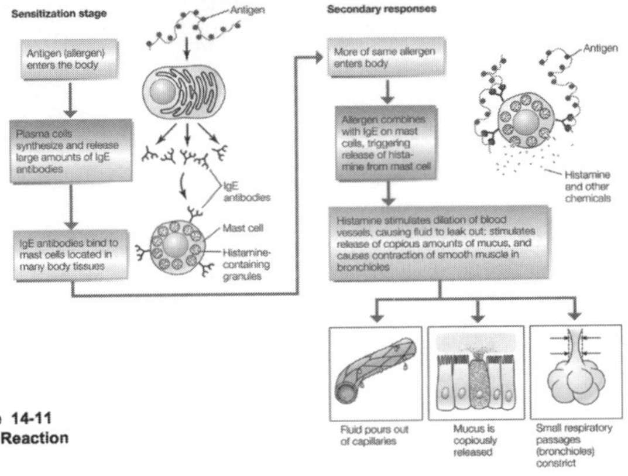

Figure 14-11
Allergic Reaction

Notes

Section 14-5 Health and Homeostasis

The immune system's protective mechanisms play a vital role in homeostasis. Without the body's defenses, potentially dangerous viruses and microorganisms may harm us.

Chapter 14A: A Closer Look: Aquired Immunodeficiency Syndrome

Notes

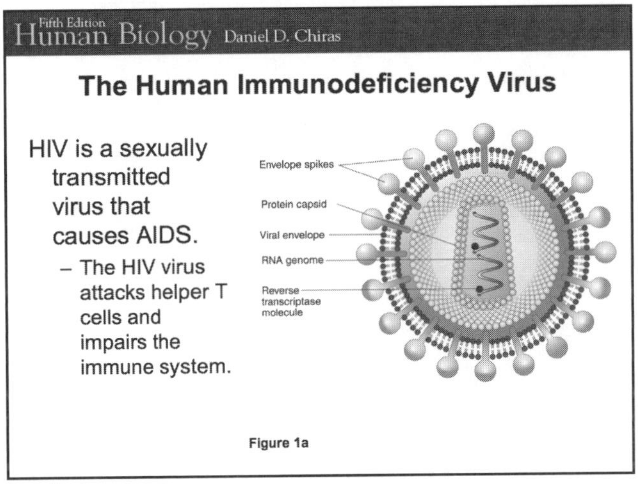

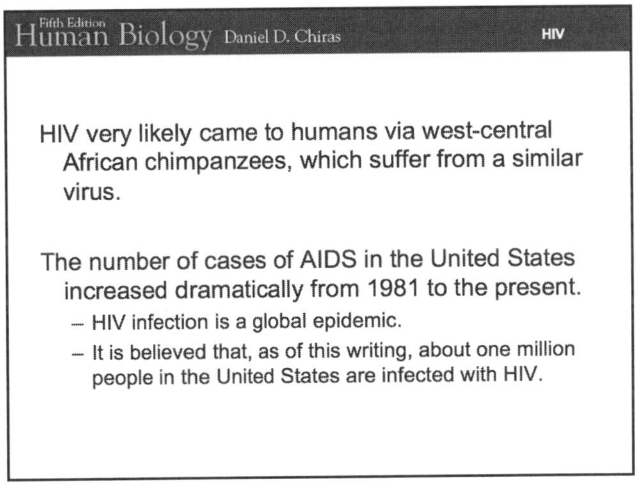

Notes

Fifth Edition Human Biology Daniel D. Chiras

Stages of AIDS

AIDS is a progressive disease that exhibits three distinct phases.

- Acute phase
- Chronic phase
- Full-blown AIDS

Fifth Edition Human Biology Daniel D. Chiras — Stages of AIDS

During the acute phase, the T4-cell count is high. Antibodies to HIV increase.

During the chronic phase, T4-cell count begins to drop in response to the increase in HIV. Outward signs of the disease begin to appear.

During full-blown AIDS, the patient has persistent infections, extreme weight loss, and weakness. Most patients perish from infections caused by microbes that are not usually life-threatening.

Fifth Edition Human Biology Daniel D. Chiras

Cancer, Brain Deterioration, and HIV

HIV also causes cancer and produces a substance that may cause deterioration of brain function.

- Kaposi's sarcoma, a rare form of skin cancer, often affects AIDS patients.

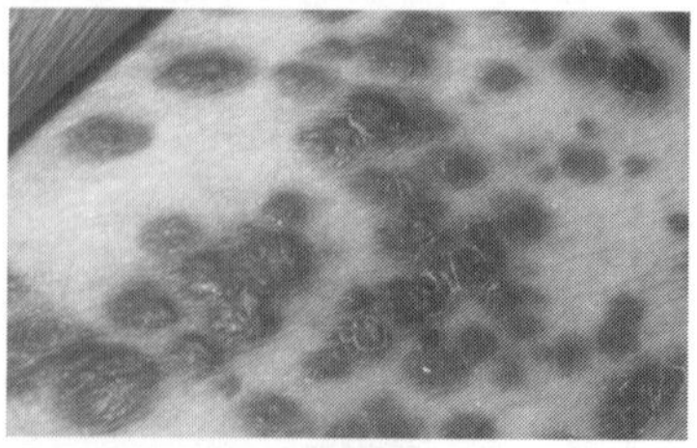

Figure 4

Courtesy of the National Cancer Institute

Fifth Edition Human Biology Daniel D. Chiras — Cancer and HIV

- In the second phase, HIV produces proteins, one of which may kill brain cells and cause memory loss and progressive mental deterioration.

Fifth Edition Human Biology Daniel D. Chiras

HIV Transmission

HIV is transmitted in many ways, but not by casual contact.

HIV is primarily passed via:

- Sexual contact
- Blood transfusions
- Contaminated needles shared by intravenous drug abusers.
- HIV can also be transmitted from mother to baby through the placenta or milk.

Fifth Edition Human Biology Daniel D. Chiras

Preventing the Spread of AIDS

AIDS can be prevented by practicing sexual abstinence, using condoms and other measures.

- Pregnant women infected with HIV can take AZT during pregnancy or try new drug treatments.

Notes

Notes

Fifth Edition Human Biology Daniel D. Chiras

Promising Development

The battle against AIDS has been facilitated by new screening tests and by drugs that slow down the development of the disease.

Although some researchers are optimistic about finding a vaccine for HIV, not all share their view.

Notes

Fifth Edition Human Biology Daniel D. Chiras

Microorganisms may cause disease in even normally healthy individuals, but those with weakened immune systems are more at risk.

Fifth Edition Human Biology Daniel D. Chiras

Section 15-1 The Infection Agents of Human Disease

Pathogens are microbes that cause disease.

Infectious microbes invade and grow in body tissues.

Contagious microbes can be easily transmitted from one organism to another.

Notes

Human Biology Fifth Edition Daniel D. Chiras — Section 15-1

Many infectious diseases remain dangerous today.

- The germ theory of disease, generated from Pasteur and Koch's work, increased our understanding of how microorganisms relate to disease.

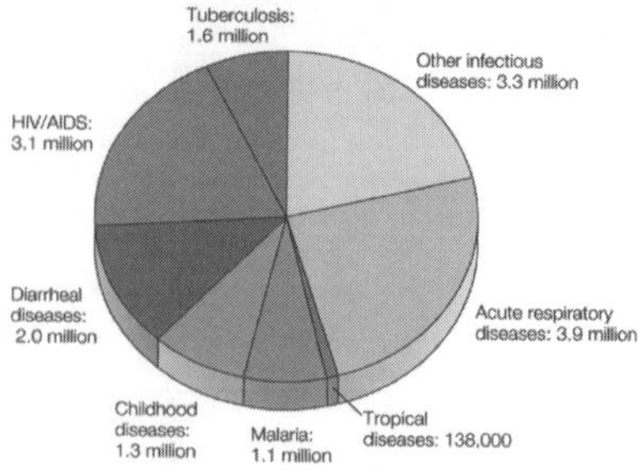

Figure 15-2

Human Biology Fifth Edition Daniel D. Chiras — Section 15-1

Viruses and bacteria are among the most common infectious agents.

Viruses are composed of

- a nucleic acid core, consisting of either DNA or RNA
- a coat of protein molecules called the capsid.
- Some capsids are covered by a viral envelope.

Human Biology Fifth Edition Daniel D. Chiras — Section 15-1

Viral life cycle

The virus

- attaches to target cells
- enters the cell through phagocytosis
- replicates its genetic information
- assembles genetic information in capsids and forms new viral particles
- are released from the host cell

Fifth Edition
Human Biology Daniel D. Chiras
Section 15-1

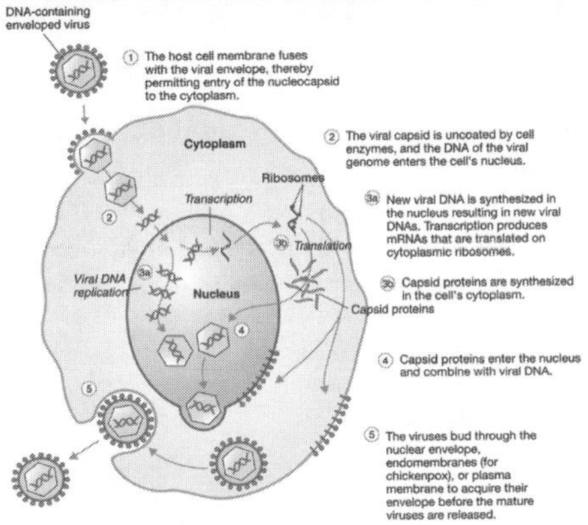

Figure 15-4 Replication of a DNA Animal Virus

Fifth Edition
Human Biology Daniel D. Chiras
Section 15-1

– Bacteria are single-celled organisms shaped like rods, spheres, or spirals. They reproduce by replicating their DNA and splitting into two identical cells.

Fifth Edition
Human Biology Daniel D. Chiras
Section 15-1

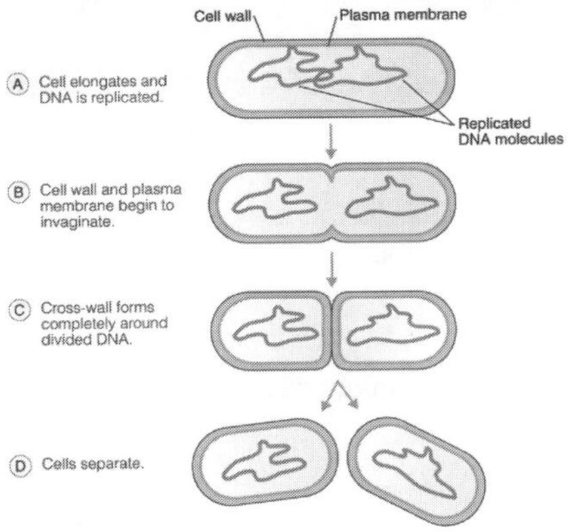

Figure 15-6a Bacterial Cell Division

Notes

Notes

Fifth Edition
Human Biology Daniel D. Chiras
Section 15-1

Fungi, protozoa, and helminths also cause disease.

- Fungi include yeasts, molds, and mushrooms.
- Protozoa are single-celled eukayotic organisms. Many harmless protozoa live in human intestines, while others invade the body and cause disease.
- Helminths are parasitic worms, such as flatworms and roundworms.

Fifth Edition
Human Biology Daniel D. Chiras
Section 15-1

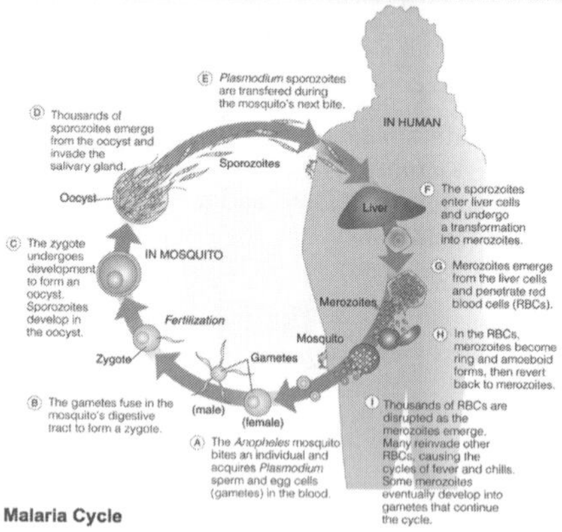

Figure 15-8 The Malaria Cycle

Fifth Edition
Human Biology Daniel D. Chiras

Section 15-2
The Course of a Human Disease

Infection refers to the state of being infected by a microorganism or to the process of being infected.

Disease is a change from the healthy state of the body.

Human Biology, Fifth Edition — Daniel D. Chiras — Section 15-2

Pathogens differ in their ability to cause disease.

- Pathogenicity is a microorganism or virus's ability to enter the host's tissues and cause a change resulting in poor health.

Human Biology, Fifth Edition — Daniel D. Chiras — Section 15-2

A pathogen produces a set of disease characteristics.

- Every disease that infects the body changes the body structures and functions in specific ways.
  - Signs are detectable and measurable changes in body function (fever, swollen lymph).
  - Symptoms are un-measurable changes in the body (headache, tiredness).
  - A syndrome is a specific group of signs and symptoms.

Human Biology, Fifth Edition — Daniel D. Chiras — Section 15-2

A disease typically follows a series of five stages.

1. Incubation period
2. Prodromal phase
3. Acme period
4. Period of decline
5. Period of convalescence

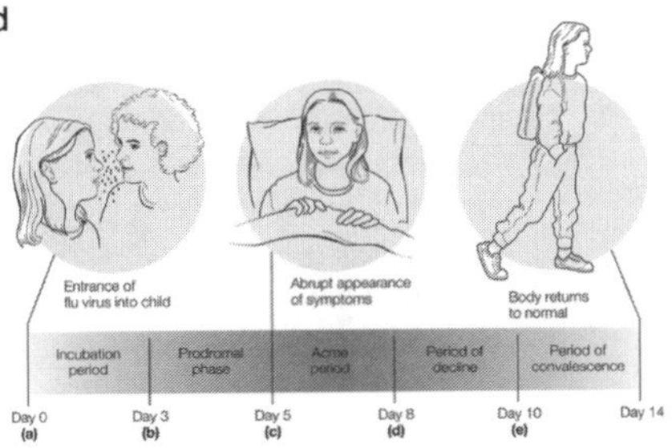

Figure 15-10

Notes

Notes

Fifth Edition Human Biology Daniel D. Chiras

Section 15-3 How Pathogens Cause Disease

Infection depends on a pathogen's ability to adhere to cells in specific tissues.

Pathogenicity and virulence depend on key metabolic characteristics.

- Ezymes and toxins released by the pathogen affect the microorganism's ability to penetrate and damage cells.

Fifth Edition Human Biology Daniel D. Chiras — Section 15-3

Toxins, microbial poisons, have two categories: exotoxins and endotoxins.

- Exotoxins are protein molecules made by bacteria present in body tissues. Even minute amounts of an exotoxin can be fatal.
- Endotoxins re a lipid portion of the cell wall of many bacteria. When the cell dies, endotoxins may be released.

Fifth Edition Human Biology Daniel D. Chiras

Section 15-4 How Infectious Diseases are Transmitted

Pathogens must be transmitted to other hosts to spread disease.

- Endemic diseases occur in a small region to a small number of individuals.
- Epidemic diseases spread a larger than normal number of individuals in a large population.
- An outbreak is a disease that occurs to a larger number of individuals within a small population.
- A pandemic disease occurs worldwide.

Fifth Edition Human Biology Daniel D. Chiras Section 15-4

Direct transmission may occur through direct physical contact.

Direct transmission method include:

- Person-to-person contact
- mother-to-child contact
- animal-to-person contact
- airborne particles and respiratory droplets

Fifth Edition Human Biology Daniel D. Chiras Section 15-4

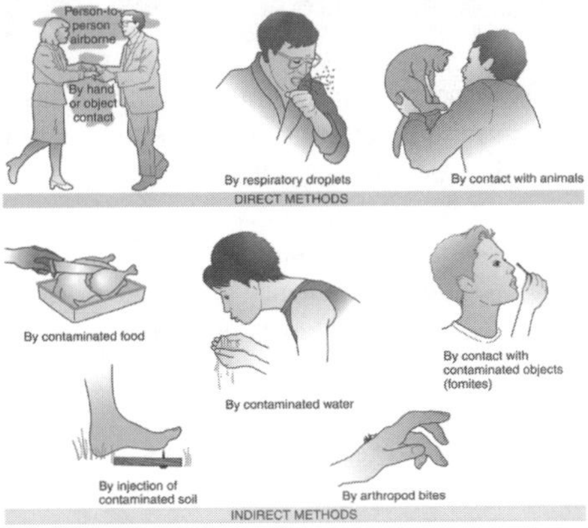

Figure 15-12 Methods of Transmitting Disease

Fifth Edition Human Biology Daniel D. Chiras Section 15-4

Diseases can be transmitted by indirect contact.

Indirect transmission includes:

- Touching inanimate objects covered with pathogens
- Breathing pathogens carried on aerosols, tiny airborne particles
- Being bitten by arthropods (mosquitoes, flies, ticks)
- Consuming contaminated food and water

Notes

Notes

Fifth Edition Human Biology Daniel D. Chiras

Section 15-5 Emerging Infectious Diseases and Bioterrorism

The emergence and resurgence of infectious diseases have several causes.

Emerging diseases are those that have recently appeared in a population.

- Expanding world population is one major cause for emerging diseases.
- Worldwide animal transport is another cause.
- Increased international travel may spread disease to new areas.
- Changes in food handling or processing may spread disease as well.

Fifth Edition Human Biology Daniel D. Chiras — Section 15-5

Reemerging infectious diseases have existed in the past but now showing an increase in frequency or geographic range.

Causes for reemerging diseases

- Antibiotic resistance
- Lowered immunity to infectious disease
- Poor public health programs

Climate change may be a cause for emerging and reemerging diseases.

Fifth Edition Human Biology Daniel D. Chiras — Section 15-5

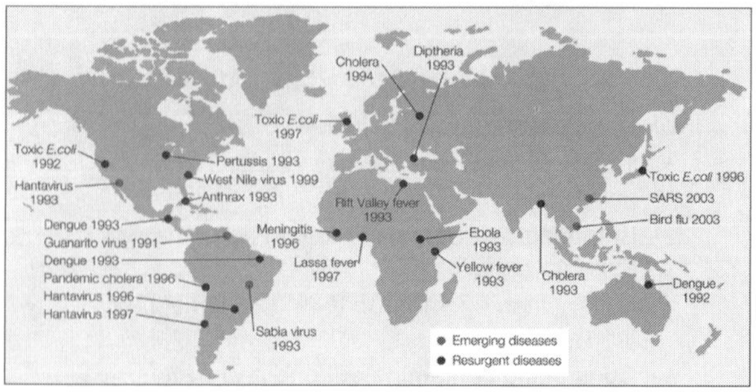

Figure 15-13 Emerging and Reemerging Diseases in the 1990s

Fifth Edition Human Biology Daniel D. Chiras — Section 15-5

Bioterrorism is an attempt to use infectious disease agents to inflict pain, suffering, and even death on large populations.

- Pathogens that cause the most concern are those that can be spread through aerosols and substances that can spread to food and water supplies.
- Developing and maintaining medical treatments is one of the best defenses against bioterrorism.

Fifth Edition Human Biology Daniel D. Chiras

Section 15-6 Health and Homeostasis

The chances of getting an emerging or reemerging infectious disease are relatively low.

Contracting a disease like *Salmonella* is much more likely.

- It is important to defend oneself against the common diseases through hand washing, proper food preparation, sanitation, and vaccination.

Notes

Notes

Chapter 16

Chromosomes, Cell Division, and the Cell Cycle

Fifth Edition Human Biology Daniel D. Chiras

Section 16-1 The Cell Cycle

The cell cycle is divided into two parts:

– Interphase
– Cell division

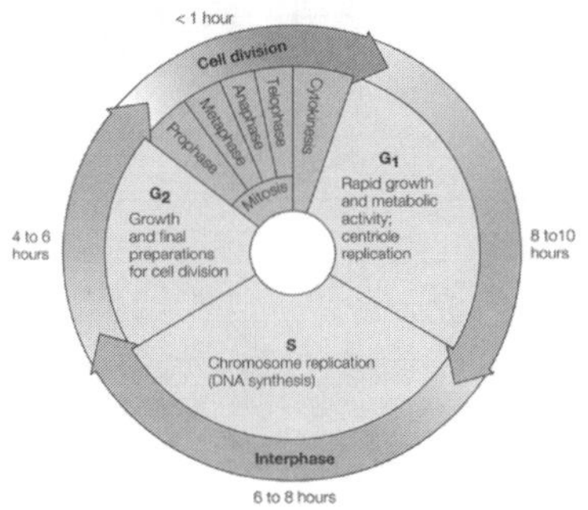

Figure 16-2

Fifth Edition Human Biology Daniel D. Chiras — Section 16-1

Interphase is divided into three parts:

– G1, S, and G2
  - During G1, the cell produces RNA, proteins, and other molecules.
  - During the S phase, the DNA replicates
  - During G2, mitochondria divide and precursors of the spindle fibers form

Notes

Fifth Edition Human Biology Daniel D. Chiras — Section 16-1

Nuclear and cytoplasmic division occur separately.

- Cell division follows interphase.
- It requires two separate but related processes: mitosis, or nuclear division, and cytokinesis, or cytoplasmic division.

Fifth Edition Human Biology Daniel D. Chiras

Section 16-2 The Chromosome

Human body cells contain 46 chromosomes that contain all of the genetic information required to control cellular activity.

- Each organism has a set number of chromosomes
- All body cells, except the germ cells, are called somatic cells.
  - Body cells contain a full complement of chromosomes and are described as diploid.
  - Germ cells or gametes contain half the number of chromosomes of somatic cells and are referred to as haploid cells.

Fifth Edition Human Biology Daniel D. Chiras — Section 16-2

Chromosomes condense after replication, which facilitates mitosis.

- Condensed chromosomes are metabolically inactive.

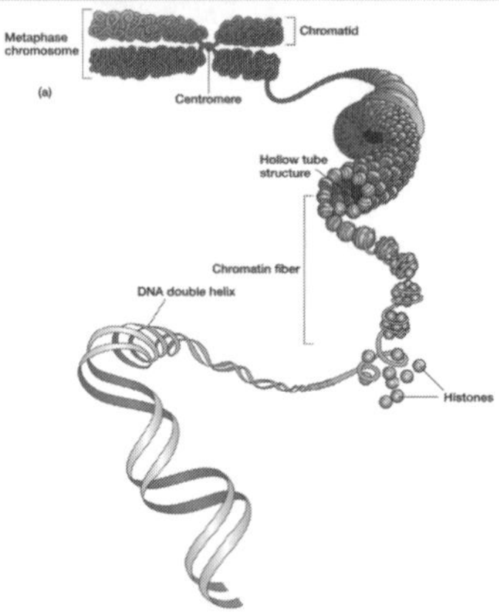

Figure 16-6a

Fifth Edition Human Biology Daniel D. Chiras Section 16-2

The number of chromatids in chromosomes varies, depending on the stage of the cell cycle.

Fifth Edition Human Biology Daniel D. Chiras

Section 16-3 Cell Division

Mitosis is divided into four stages:
- Prophase
- Metaphase
- Anaphase
- Telophase

Fifth Edition Human Biology Daniel D. Chiras Section 16-3

Prophase
- Chromosomes condense
- Nuclear envelope disappears
- Centrioles divide and migrate to opposite poles of the dividing cell
- Spindle fibers form and attach to chromosomes

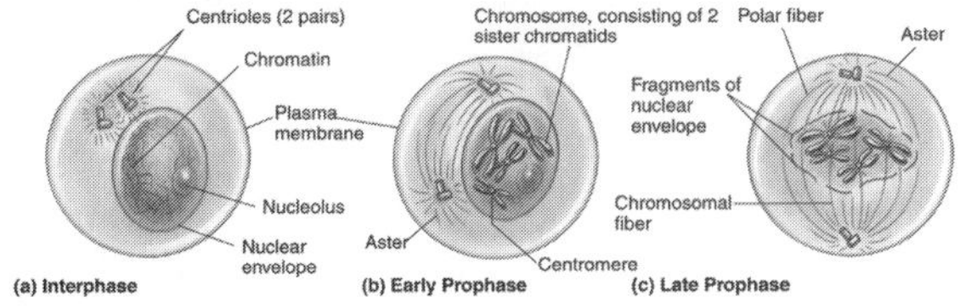

Figure 16-8a-c

Notes

Notes

Fifth Edition Human Biology Daniel D. Chiras — Section 16-3

Metaphase

- Chromosomes line up on equatorial plate of the dividing cell.

Anaphase

- Chromosomes begin to separate.

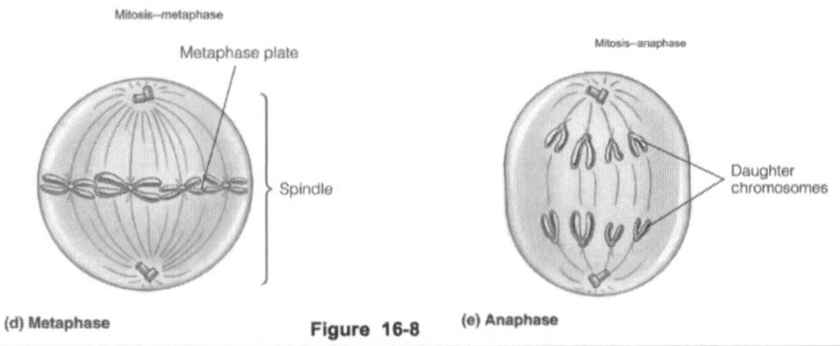

Figure 16-8

Fifth Edition Human Biology Daniel D. Chiras — Section 16-3

Telophase

- Chromosomes migrate or are pulled to opposite poles.
- New nuclear envelope forms.
- Chromosomes uncoil.

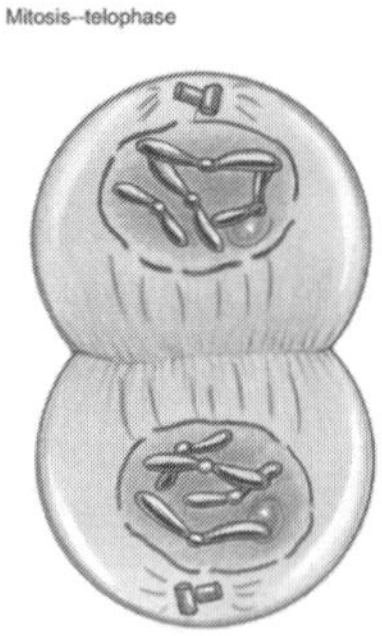

(f) Telophase and cytokinesis

Figure 16-8

Fifth Edition Human Biology Daniel D. Chiras — Section 16-3

Cytokinesis, the division of the cytoplasm, is made possible by contraction of microfilaments beneath the cell membrane.

- Cytokinesis begins in late anaphase or early telophase.
- In human cells, cytokinesis results from the contraction of microfilaments lying beneath the plasma membrane.

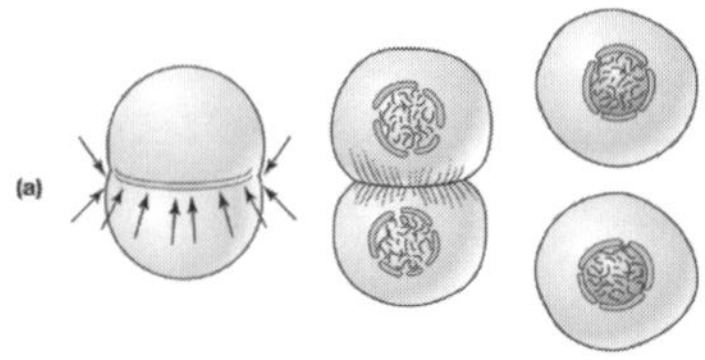

Figure 16-11

Fifth Edition Human Biology Daniel D. Chiras

Section 16-4 Control of the Cell Cycle

Many factors appear to control the cell cycle.

- Research suggests the cell cycle is controlled in part by chemical messages produced in the cytoplasm.
- External controls such as hormones, growth regulators, and cell contact are imposed on the cell.

Notes

Notes

Chapter 17

Principles of Human Heredity

Fifth Edition Human Biology Daniel D. Chiras

Section 17-1 Meiosis and Gamete Formation

During gamete formation, germ cells undergo a special kind of division called meiosis, which reduces the number of chromosomes by half.

- Meiosis is a type of nuclear division found only in germ-cell production in the gonads of sexually reproducing animals.

Fifth Edition Human Biology Daniel D. Chiras — Section 17-1

Meiosis involves two cellular divisions.

Meiosis I is a reduction division.

- The chromosome number is halved.

Meiosis II is like mitosis in many ways.

- When haploid cells divide in meiosis II, they produce two new cells.
- Each cell contains a haploid number of single-stranded chromosomes.

Notes

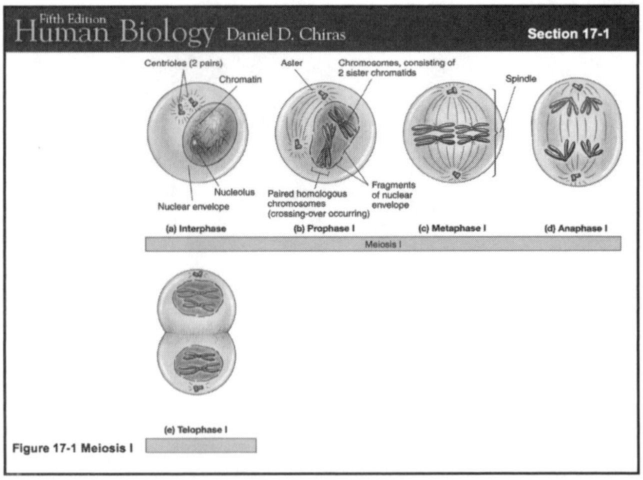

Figure 17-1 Meiosis I

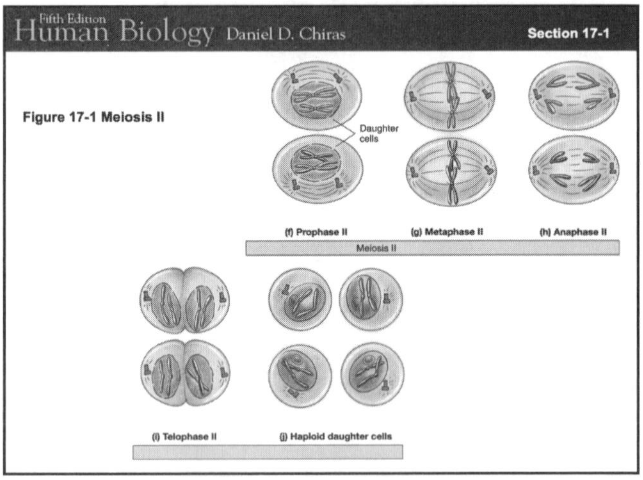

Figure 17-1 Meiosis II

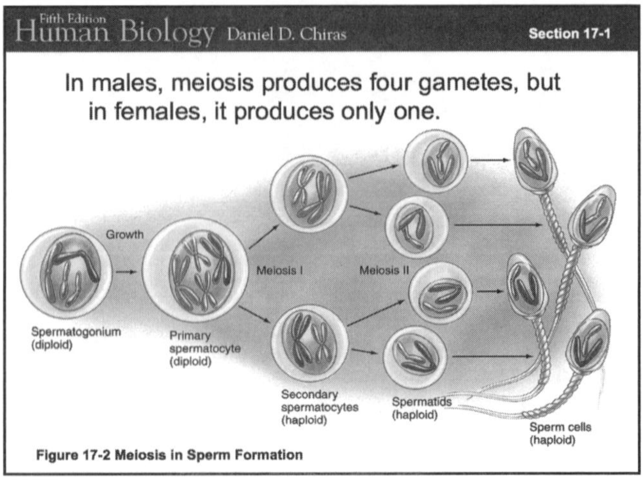

Figure 17-2 Meiosis in Sperm Formation

Fifth Edition Human Biology Daniel D. Chiras — Section 17-1

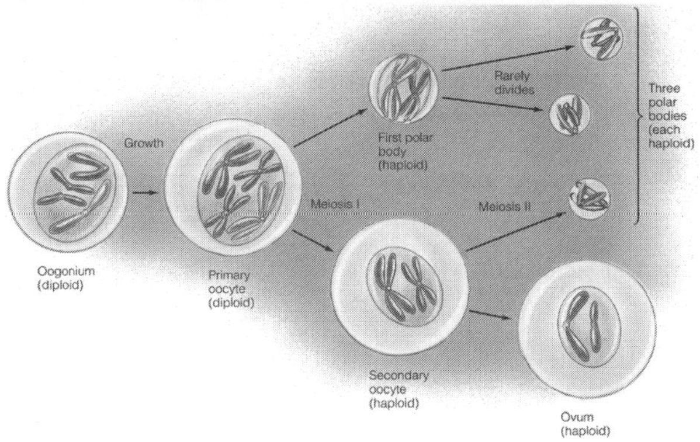

Figure 17-3 Meiosis in Ovum Formation

Fifth Edition Human Biology Daniel D. Chiras

Section 17-2 Principles of Heredity

Much of our early knowledge on heredity came from the work of Gregor Mendel.

Mendel discovered that the traits he was studying did not blend.

Mendel discovered that the parents contributed equally to the characteristics of their offspring.

Fifth Edition Human Biology Daniel D. Chiras — Section 17-2

Mendel also discovered the principle of dominance.
- A dominant factor masks a recessive factor.
- A recessive factor is expressed only when the dominant factor is missing.
- The dominant and recessive genes are alternative forms of the gene, or alleles.
- Three genetic combinations are possible for a given trait:
  - Heterozygous
  - Homozygous dominant
  - Homozygous recessive

Notes

Notes

Fifth Edition Human Biology Daniel D. Chiras — Section 17-2

The genotype of an organism is its genetic makeup; the phenotype is its appearance.

Genotypes and phenotypes can be determined by the Punnett square.

Fifth Edition Human Biology Daniel D. Chiras — Section 17-2

Genes on different chromosomes segregate independently of one another during gamete formation.

- This is the principle of independent assortment. It holds true only for nonlinked genes.

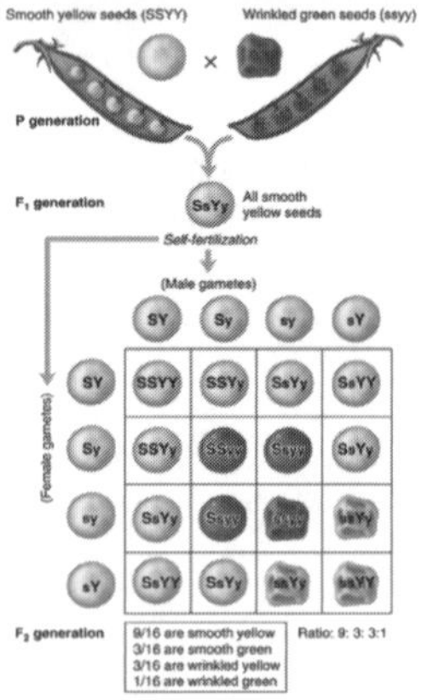

Figure 17-7

Fifth Edition Human Biology Daniel D. Chiras

Section 17-3 Mendelian Genetics in Humans

Autosomal-recessive traits are expressed only when both alleles are recessive.

- Human cells contain 23 pairs of chromosomes:
  - 22 pairs of autosomes and 1 pair of sex chromosomes.
- Chromosomes carry dominant and recessive traits.
- Inheritance of these traits is consistent with Mendel's principles of inheritance.

Autosomal-dominant traits are expressed in all individuals.

Fifth Edition
Human Biology Daniel D. Chiras

Section 17-4 Variations in Mendelian Genetics

Incomplete dominance results in a kind of blending of traits, called incomplete, or partial, dominance.

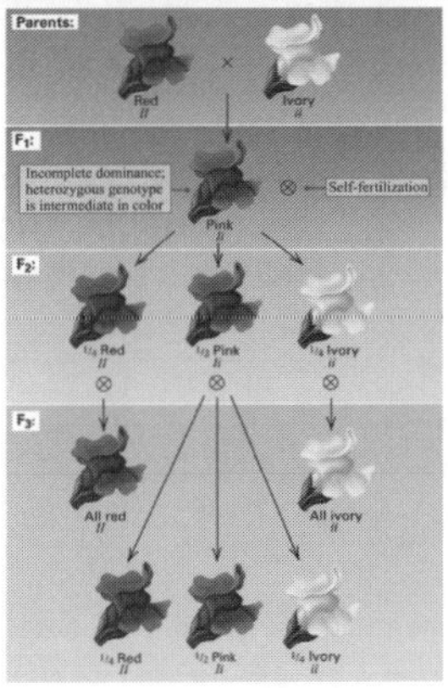

Figure 17-12

Fifth Edition
Human Biology Daniel D. Chiras — Section 17-4

Some genes have multiple alleles.

- Multiple alleles result in more genotypes and phenotypes in a population.
- Because chromosomes exist in pairs, individuals can have only two of the possible alleles.
- Codominance occurs in multiple-allele genes. Codominant genes are expressed fully and equally.

Fifth Edition
Human Biology Daniel D. Chiras — Section 17-4

Some traits are determined by more than one gene pair.

- This phenomenon is referred to as polygenic inheritance.

Crossing over is a genetic swapping that occurs in meiosis.

- If crossing-over does not occur, these linked genes are inherited together.

Notes

Notes

Human Biology, Fifth Edition, Daniel D. Chiras — Section 17-4

The Human Genome Project seeks to determine the sequence of bases in human DNA and the location of all of the genes.

Human Biology, Fifth Edition, Daniel D. Chiras

Section 17-5 Sex-Linked Genes

The presence or absence of the Y chromosome determines one's sex.

If it is absent, the individual (XX) is a female.

If it is present, the individual (XY) is a male.

Human Biology, Fifth Edition, Daniel D. Chiras — Section 17-5

– The sex chromosomes also carry genes that determine physical traits.

- A trait determined by a gene on a sex chromosome is a sex-linked trait.
- Most sex-linked traits occur on the X chromosome; these are called x-linked genes.

Notes

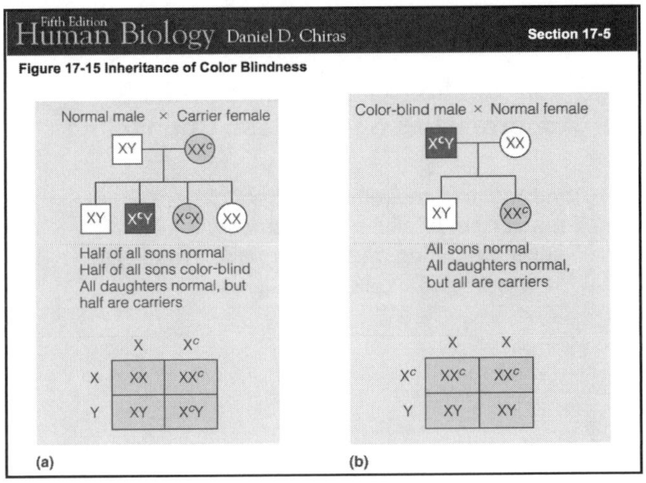

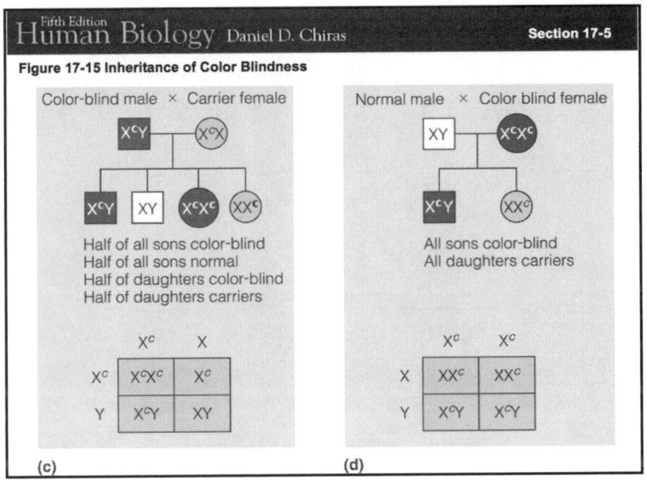

Fifth Edition Human Biology Daniel D. Chiras — Section 17-5

In order for a recessive gene to be expressed, it must be located on both X chromosomes in females, but only one in males.

When present, a dominant gene is expressed in males and females.

- Recessive X-linked genes are the best understood of the sex-linked genes.
- Dominant X-linked genes are relatively rare.

Both dominant and recessive Y-linked genes are always expressed.

The action of some genes is influenced by the sex of an individual.

Notes

Fifth Edition
Human Biology Daniel D. Chiras

Section 17-6 Chromosomal Abnormalities and Genetic Counseling

Abnormal chromosome numbers generally result from a failure of chromosomes to separate during gamete formation (meiosis).
- This process is called nondisjunction.

Fifth Edition
Human Biology Daniel D. Chiras

Section 17-6

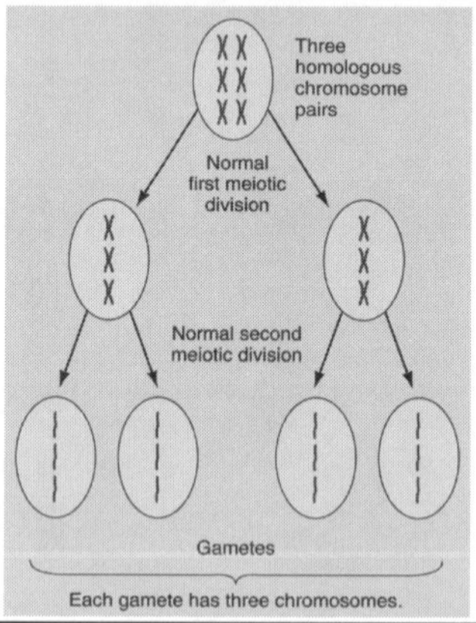

Figure 17-18a Meiosis and Abnormal Chromsome Numbers

Fifth Edition
Human Biology Daniel D. Chiras

Section 17-6

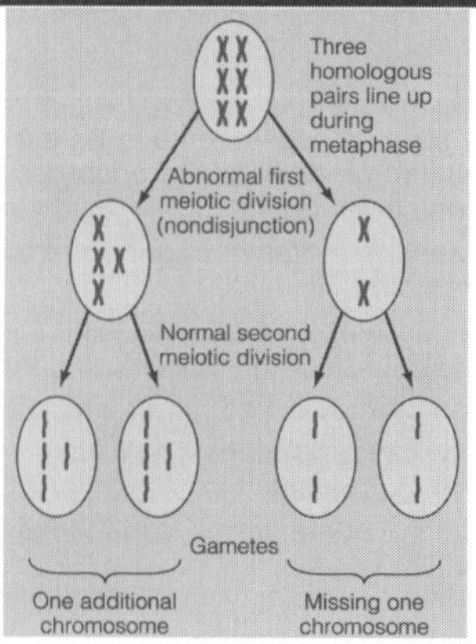

Figure 17-18b Meiosis and Abnormal Chromsome Numbers

Fifth Edition Human Biology Daniel D. Chiras — Section 17-6

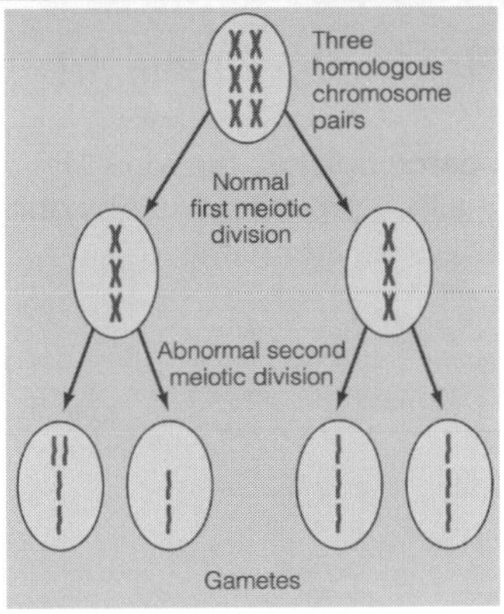

Figure 17-18c Meiosis and Abnormal Chromsome Numbers

Fifth Edition Human Biology Daniel D. Chiras — Section 17-6

Genetic disorders may also result from variations in chromosome structure.

- Variations result from two occurrences:
  - Deletions—the loss of a piece of chromosome
  - Translocations—breakage followed by reattachment elsewhere

Fifth Edition Human Biology Daniel D. Chiras — Section 17-6

Genetic screening allows parents to determine whether they will have a genetically normal baby.

DNA abnormalities may occur in mitochondria, resulting in rare diseases.

Notes

Notes

Fifth Edition Human Biology Daniel D. Chiras

Section 17-7 Health and Homeostasis

New research bolsters the idea that our personalities are influenced by our genes as well as our environment.

Notes

Chapter 18

How Genes Work and How Genes are Controlled

Fifth Edition Human Biology Daniel D. Chiras

Section 18-1 DNA and RNA: Macromolecules with a Mission

DNA consists of a double helix held together by hydrogen bonds.

- Each strand of the double helix contains nucleotides.
- Each nucleotide in the DNA molecule consists of a purine or pyrimidine base, the sugar deoxyribose, and a phosphate group.
- The nucleotides are joined by covalent bonds between phosphate groups and deoxyribose molecules.

Fifth Edition Human Biology Daniel D. Chiras — Section 18-1

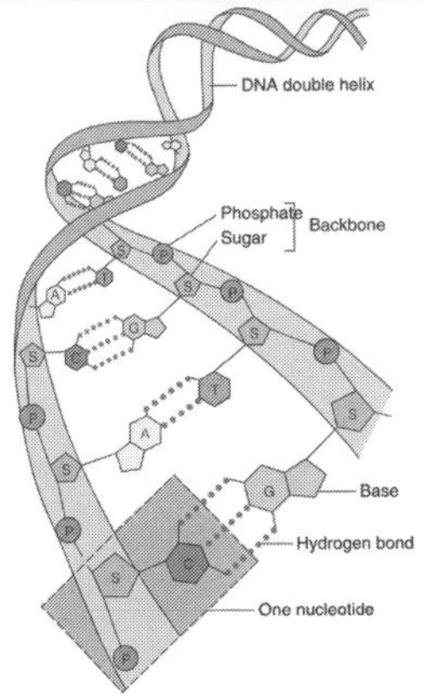

Figure 18-1 DNA

Notes

Fifth Edition Human Biology Daniel D. Chiras — Section 18-1

- Complementary base pairing is an unalterable coupling in which adenine on one strand of the DNA molecule always binds to thymine on the other and guanine always binds to cytosine.
- Complementary base pairing ensures
  - accurate replication of the DNA
  - accurate transmission of genetic information from one cell to another and from one generation to another.

Fifth Edition Human Biology Daniel D. Chiras — Section 18-1

DNA unwinds then serves as a template for the production of new DNA strands.

- DNA polymerase is an enzyme that helps align and pair nucleotides to the template strand.

Fifth Edition Human Biology Daniel D. Chiras — Section 18-1

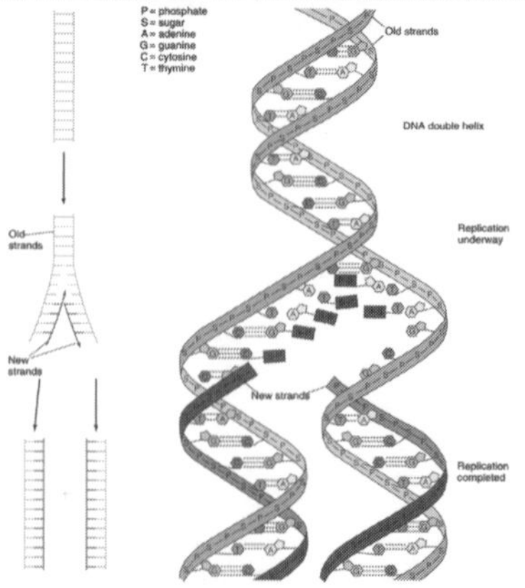

Figure 18-3 DNA Replication

Fifth Edition Human Biology Daniel D. Chiras

Section 18-1

Three types of RNA exist, each of which is involved in protein synthesis:

- Transfer RNA
- Ribosomal RNA
- Messenger RNA.

All three RNA molecules are single-stranded polynucleotide chains.

Fifth Edition Human Biology Daniel D. Chiras

Section 18-1

RNA synthesis is called transcription and takes place on a DNA template in the nucleus of the cell.

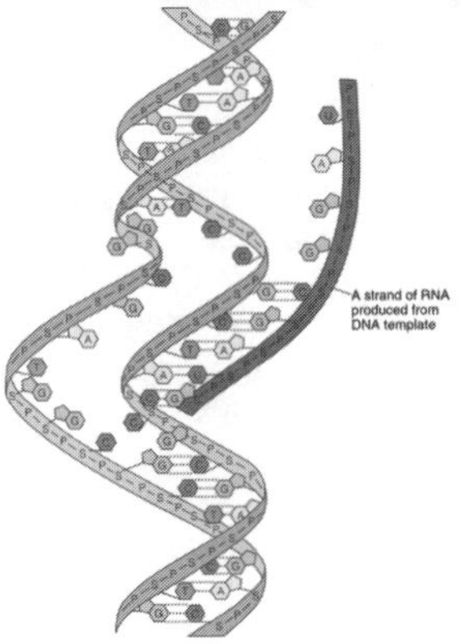

Figure 18-6

Fifth Edition Human Biology Daniel D. Chiras

Section 18-2 How Genes Work: Protein Synthesis

Protein is synthesized on a mRNA template.

- This process is called translation.
- The genetic information contained in the DNA molecule is transferred to messenger RNA.
- Messenger RNA molecules carry this information to the cytoplasm, where proteins are synthesized.
- Messenger RNA serves as a template for protein synthesis.
- Ribosomes are required to produce proteins on the mRNA template.

Notes

Notes

Fifth Edition Human Biology Daniel D. Chiras — Section 18-2

- Transfer RNA molecules deliver amino acid molecules to the mRNA and insert them in the growing chain.
- Each tRNA binds to a specific amino acid and delivers it to a specific codon, a sequence of three bases on the mRNA.
- The sequence of codons determines the sequence of amino acids in the protein.
- Messenger RNA serves as a template for protein synthesis.

Fifth Edition Human Biology Daniel D. Chiras — Section 18-2

- Proteins are synthesized by adding one amino acid at a time.
- During protein synthesis, the ribosome first attaches to the mRNA at the initiator codon.
- Soon after, the large subunit attaches.
- A specific tRNA bound to an amino acid binds to the initiator codon and the first binding site of the ribosome.
- A second tRNA–amino acid then enters the second site.

Fifth Edition Human Biology Daniel D. Chiras — Section 18-2

- An enzyme in the ribosome catalyzes the formation of a peptide bond between the two amino acids.
- After the bond is formed, the first tRNA (minus its amino acid) leaves the first binding site.

Fifth Edition Human Biology Daniel D. Chiras — Section 18-2

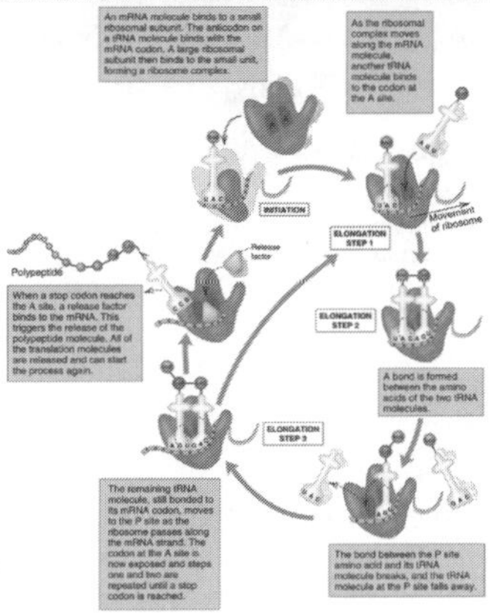

Figure 18-10 Protein Synthesis

Fifth Edition Human Biology Daniel D. Chiras — Section 18-2

- The ribosome moves down the mRNA, shifting the tRNA bound to its two amino acids to the first binding site and opening the second site for another tRNA–amino acid.
- This process repeats itself many times in rapid succession.

Fifth Edition Human Biology Daniel D. Chiras — Section 18-2

As the peptide chain is formed, hydrogen bonds begin to form between the amino acids, and the chain begins to bend and twist, forming the secondary structure of the protein or peptide. When the ribosome reaches the terminator codon, the peptide chain is released.

Notes

Notes

Fifth Edition Human Biology Daniel D. Chiras

Section 18-3 Controlling Gene Expression

In humans, genetic expression is controlled at four levels:

- At the chromosome—access to the genes is controlled by coiling and uncoiling of the chromosome during interphase.
- At transcription—three control mechanisms operate at the level of transcription:
  - Induction
  - Repression
  - Enhancement
- After transcription but before translation—by altering the structure of mRNA.
- At translation—by masking mRNA

Fifth Edition Human Biology Daniel D. Chiras

Section 18-3

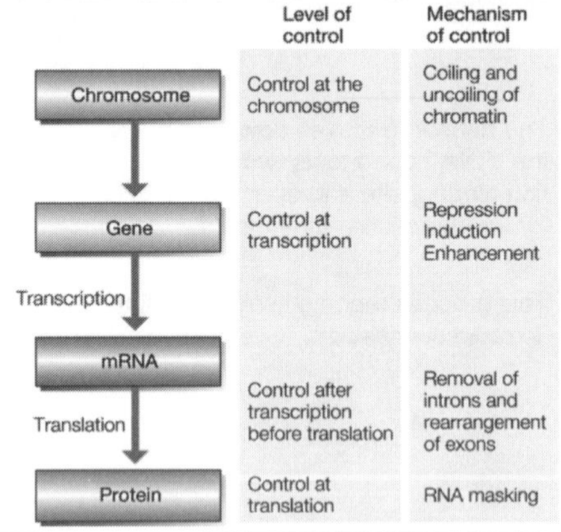

Figure 18-12 Gene Expression

Fifth Edition Human Biology Daniel D. Chiras

Section 18-4 Health and Homeostasis

Humans and other organisms contain proto-oncogenes, which control functions related to cellular replication.

Mutations in these genes caused by chemical, physical, biological agents, or viruses may cause cancer.

Chapter 19: Genetic Engineering and Biotechnology

Notes

Fifth Edition Human Biology Daniel D. Chiras

Section 19-1 Genetic Manipulation: An Overview

Genetic engineering, or recombinant DNA technology, is a procedure by which geneticists remove segments of DNA from one organism and insert them into the DNA of another.

Fifth Edition Human Biology Daniel D. Chiras

Section 19-2 Cloning

Cloning offers us an opportunity to produce identical replicas of existing or extinct organisms, using DNA from one organism and cells of another.

Cloning also refers to the replication of genes via various methods.

Notes

Section 19-3 Recombinant DNA Technology

In recombinant DNA technology, genes are removed from one organism, then spliced into another of the same or a different species.

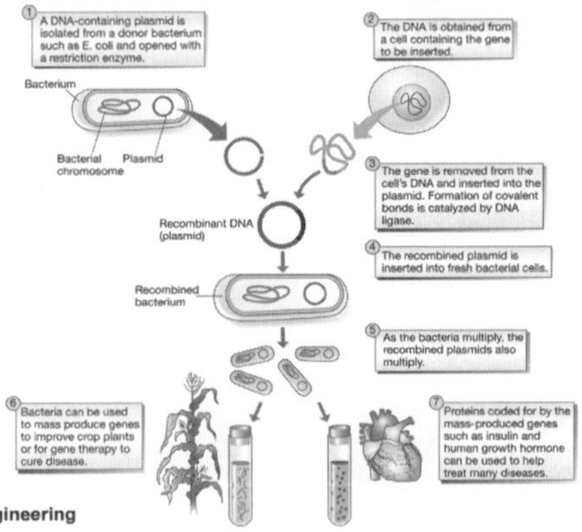

Figure 19-5 Genetic Engineering Simplified

Enzymes are used to excise segments of genes and insert them into foreign DNA molecules.

- After a segment of DNA is transplanted, a second enzyme, DNA ligase, is used to seal it in place.
- At this point, geneticists have formed a recombinant DNA molecule—a molecule that contains DNA from two different organisms.

Notes

Fifth Edition Human Biology Daniel D. Chiras — Section 19-3

Plasmids from bacteria can be used to clone DNA fragments for mass production.
- Foreign genes can be spliced into plasmids.
- The plasmid carrying a foreign gene can be reinserted into bacteria or yeast cells.
- In culture, the plasmids are rapidly taken up by the new hosts. As these cells divide, the plasmids replicate.
- This produces multiple copies of the gene.

Fifth Edition Human Biology Daniel D. Chiras — Section 19-3

Genes can be mass produced on strands of mRNA.
- Messenger RNA is extracted from cells and used to make DNA; this is called reverse transcription.

Genes can also be mass produced using the polymerase chain reaction.
- Part of the DNA double helix is extracted, heated to split the strands, then cooled. Enzymes are added to catalyze new DNA formulation.

Fifth Edition Human Biology Daniel D. Chiras

Section 19-4 Applications of Recombinant DNA Technology

Mass produced hormones and other proteins can be used to treat a many diseases.
- Recombinant DNA techniques can be used to produce vaccines.
- Gene splicing can be used to transfer genes from one organism to another.
- Transgenic plants are easier to produce than transgenic animals.
- Transgenic microorganisms are perhaps some of the most controversial of all applications.

Notes

Fifth Edition Human Biology Daniel D. Chiras

Section 19-4

- Recombinant DNA technology can be used to cure genetic disease, a treatment called gene therapy.
- DNA probes permit analysis of DNA from crime scenes, detection of infectious organisms, and genetic disorders.
- DNA probes can even be used to help geneticists map human chromosomes.

Fifth Edition Human Biology Daniel D. Chiras

Section 19-4

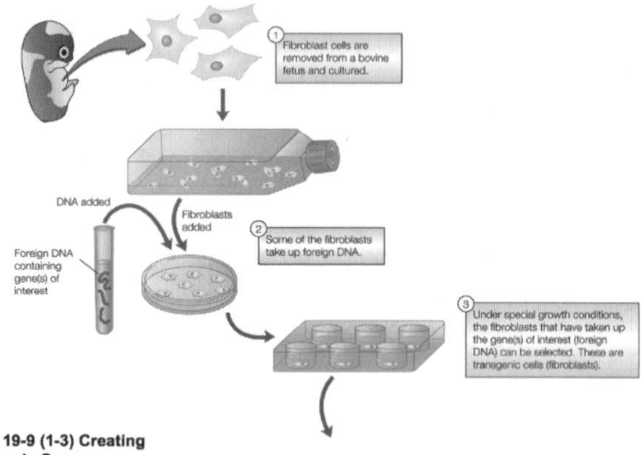

Figure 19-9 (1-3) Creating Transgenic Cows

Fifth Edition Human Biology Daniel D. Chiras

Section 19-4

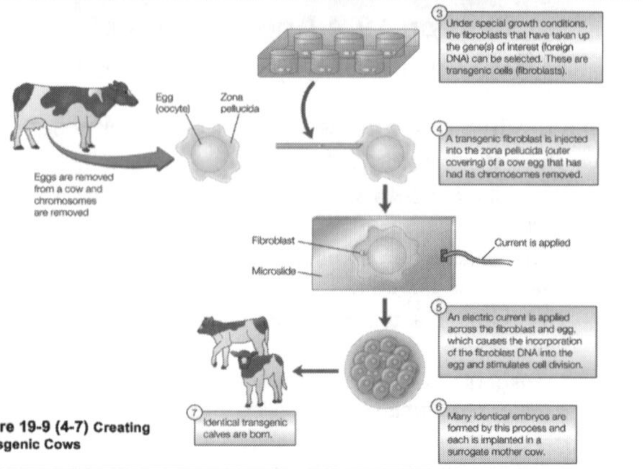

Figure 19-9 (4-7) Creating Transgenic Cows

Fifth Edition Human Biology Daniel D. Chiras

Section 19-5 Controversies over Genetic Engineering—Ethics and Safety Concerns

Despite its promises, genetic engineering is fraught with controversy over safety and ethics.

- Critics are concerned that genetically altered organisms, especially potentially dangerous bacteria or viruses might be released into the environment.
- Other critics object to what they see as tinkering with evolution.
- Experience with genetic engineering is too limited to completely answer the critics' safety concerns.

Notes

Notes

Chapter 20

Cancer

Fifth Edition Human Biology Daniel D. Chiras

Section 20-1 Benign and Malignant Tumors

Abnormal cell division may lead to benign tumors, which stop growing, or malignant tumors, which can invade other parts of the body.

Two types of tumors are encountered in humans:

- Benign tumors fail to grow and spread and rarely cause medical problems.
- Malignant tumors continue to grow, often spreading to other parts of the body and invading vital organs.
- The term *cancer* is reserved for malignant tumors.

Fifth Edition Human Biology Daniel D. Chiras — Section 20-1

Cancer cells often migrate to other sites.

- Cancer cells break loose from their site of origin, the primary tumor, and form secondary tumors in other parts of the body. Metastasis, the movement of cancerous cells from one part of the body to another, occurs most commonly in the lungs.

Notes

Fifth Edition Human Biology Daniel D. Chiras — Section 20-1

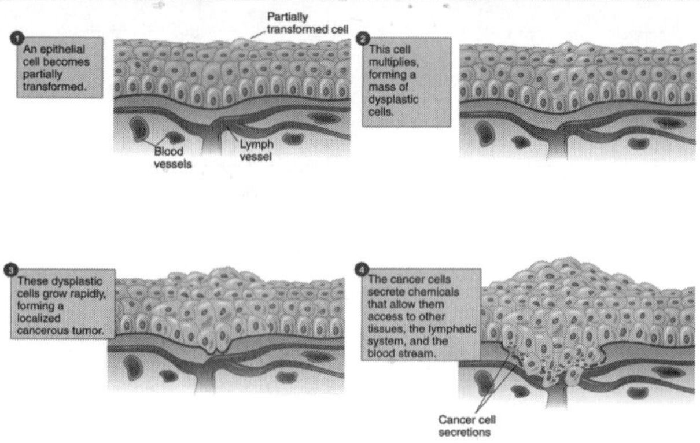

Figure 20-2 Cancer Growth and Metastasis

Fifth Edition Human Biology Daniel D. Chiras

Section 20-2 How Cancers Form

The conversion of a normal cell to a cancerous one involves two steps:

- Conversion—A mutation, caused by a chemical, biological, or physical agent, changes the DNA.
- Development and progression—A promoter stimulates the cell to proliferate uncontrollably.

Fifth Edition Human Biology Daniel D. Chiras — Section 20-2

Mutations in two groups of genes are primarily responsible for cancer.

- Proto-oncogenes control functions related to cell replication.
- Tumor suppressor genes naturally inhibit cell growth.

In the development and progression stage, chemicals called promoters stimulate mutated cells to divide uncontrollably.

Fifth Edition Human Biology Daniel D. Chiras **Section 20-2**

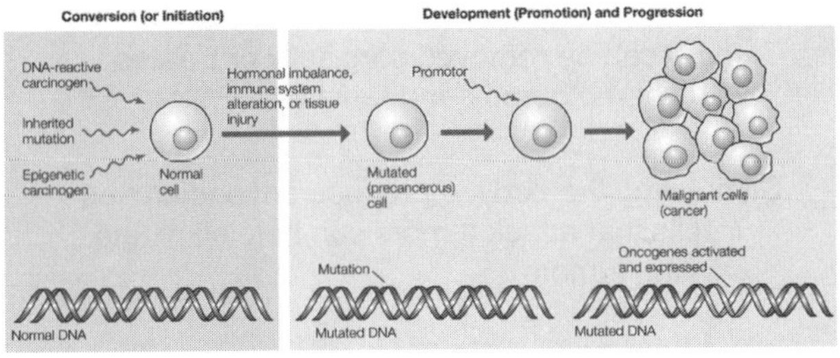

Figure 20-3 Stages Leading to Cancer

Notes

Fifth Edition Human Biology Daniel D. Chiras **Section 20-2**

Several physical and biological agents, such as X-rays, ultraviolet radiation, and viruses, are also responsible for producing cancer.

Some carcinogens exert their effect outside of the DNA.

- DNA-reactive carcinogens cause cancer by altering cell DNA.
- Epigenetic carcinogens cause changes outside the DNA that may lead to cancer.

Fifth Edition Human Biology Daniel D. Chiras **Section 20-2**

Cancers develop many years after the initial exposure to a carcinogen.

- The period between exposure and emergence of a cancer is called the latent period.

Notes

Fifth Edition Human Biology Daniel D. Chiras

Section 20-3 Cancer Treatment and Cures

Tumors can be removed surgically or treated with chemotherapy and radiation.

Stimulating the body to produce antibodies and T cells that attack tumors can help eliminate existing tumors.

Fifth Edition Human Biology Daniel D. Chiras Section 20-3

Liposomes containing cancer-killing chemicals or monoclonal antibodies attached to chemotherapeutic agents can deliver a lethal dose of these chemicals directly to cancers.

Reducing blood vessel development in tumors successfully stops cancers from growing.

Fifth Edition Human Biology Daniel D. Chiras Section 20-3

Blocking genes involved in cell replication or alternating their products, could help doctors kill cancer cells.

Drugs combined with light therapy may prove to be an effective treatment in some forms of cancer.

Fifth Edition Human Biology Daniel D. Chiras — Section 20-3

The earlier a cancer is detected and treated, the better one's chances of survival.

Preventing cancer may be the best approach of any.

Notes

Notes

Chapter 21

Human Reproduction

Fifth Edition Human Biology Daniel D. Chiras

Section 21-1 The Male Reproductive System

The male reproductive consists of seven basic components:

- testes,
- epididymis,
- vasa deferentia,
- sex accessory glands,
- urethra,
- penis, and
- scrotum.

Fifth Edition Human Biology Daniel D. Chiras — Section 21-1

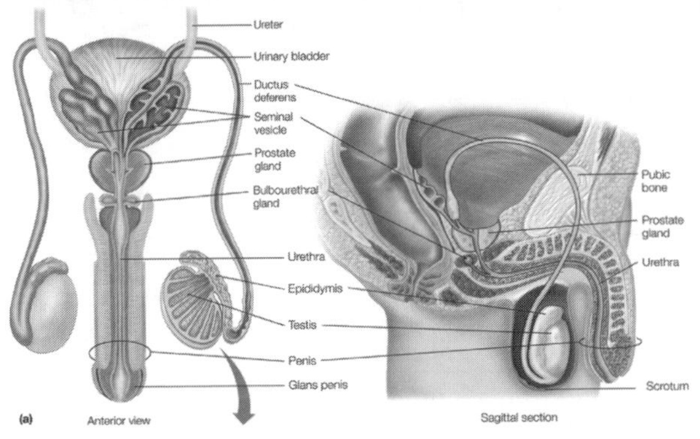

Figure 21-1a The Male Reproductive System

Notes

Sperm are produced in the seminiferous tubules and stored in the epididymis until ejaculation.

- Testes contain seminiferous tubules where sperm is formed.
- Sperm empties into a network of tubules, which empty into the epdidymal duct.
- The epididmal duct forms the epididymis.
- During ejaculation sperm moves through the epididymal duct to the vas deferens to the urethra, where semen exits the body

Fifth Edition Human Biology Daniel D. Chiras Section 21-1

Semen consists mostly of fluids produced by the sex accessory glands.

- The sex accessory glands add fluids to sperm during ejaculation. Together sperm and the fluid is called semen.

Fifth Edition Human Biology Daniel D. Chiras Section 21-1

The prostate enlarges with age and is the site of a common form of cancer in men.

Sperm are formed in the seminiferous tubules.

- Sperm are formed from special cells called spermatogonia.
- Spermatogonia that undergo meiosis yield four cells called spermatids.
- Spematids differentiate into sperm with a flagellum and an acrosome.
- The acrosome helps digests coatings surrounding the ovum.

Fifth Edition
Human Biology Daniel D. Chiras

Section 21-1

Interstitial cells produce the male sex steroid testosterone.

Testosterone stimulates sperm formation and many other functions including:

- facial hair growth, thickening of the vocal cords, laryngeal cartilage growth, sebaceous gland secretion, and bone and muscle development.

Notes

Fifth Edition
Human Biology Daniel D. Chiras

Section 21-1

Testosterone secretion may lead to acne.

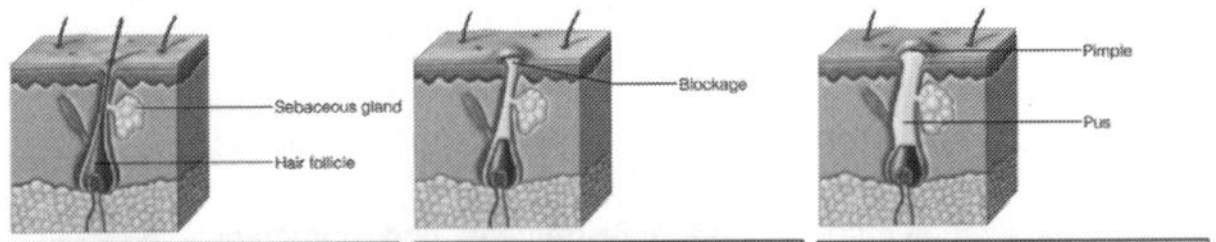

(a) Sebaceous glands associated with hair follicles secrete sebum, an oily substance that lubricates the skin.

(b) A follicle may become blocked by excess sebum and dead skin cells. Unable to escape, the sebum builds up in the hair follicle.

(c) Bacteria present on the skin may infect the sebum, causing inflammation, pus and swelling form an acne pimple.

Figure 21-4

Fifth Edition
Human Biology Daniel D. Chiras

Section 21-1

The penis contains erectile tissue that fills with blood during sexual arousal.

Ejaculation is a reflex mechanism.

Notes

Fifth Edition Human Biology Daniel D. Chiras — Section 21-1

The male reproductive system is controlled by three hormones:
- Luteinizing hormone (LH) or interstitial cell stimulating hormone (ICSH), which controls testosterone secretion
- Gonadotropin releasing hormone (GnRH), which stimulates ICSH production and release
- Follicle-stimulating hormone (FSH), which stimulates sperm formation

Fifth Edition Human Biology Daniel D. Chiras

Section 21-2 The Female Reproductive System

The female reproductive system consists of two basic components:
- The reproductive tract
- The external genitalia

The reproductive tract consists of the ovaries, uterus, oviducts, and vagina.
- The ovaries produce the female gametes and reproductive hormones.
- The uterus houses and nourishes a fetus.
- Oviducts, or uterine tubes, moves an egg from the ovaries to the uterus.
- The vagina receives sperm during sexual intercourse and expands during childbirth to allow the baby to pass.

Fifth Edition Human Biology Daniel D. Chiras — Section 21-2

The external genitalia are the externally visible parts of the female reproductive system.
- The vulva consists of two flaps of skin on both sides of the vaginal opening, the labia majora and the labia minora.
- The clitoris consists of erectile tissue and becomes filled with blood during sexual stimulation.

Many changes occur in women during sexual arousal and orgasm.

Fifth Edition Human Biology Daniel D. Chiras

Section 21-2

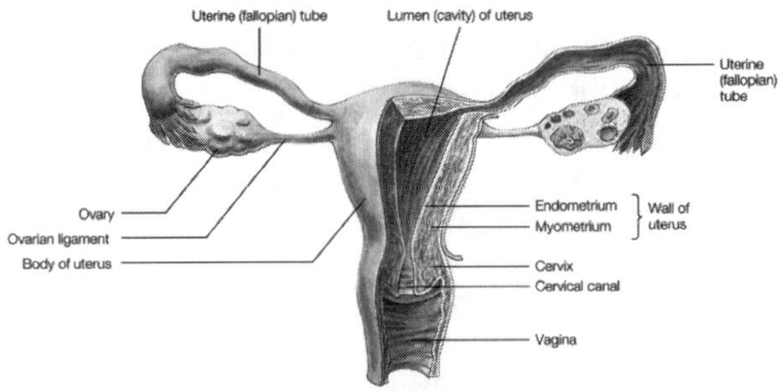

Figure 21-7a Anatomy of the Female Reproductive Tract – Frontal View

Fifth Edition Human Biology Daniel D. Chiras

Section 21-2

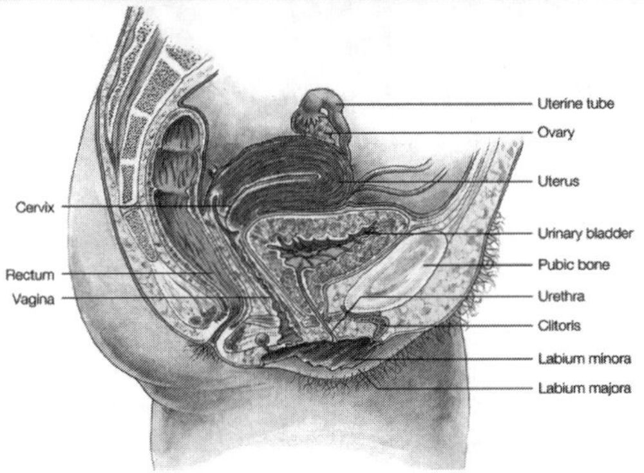

Figure 21-7b Anatomy of the Female Reproductive Tract – Midsagittal View

Fifth Edition Human Biology Daniel D. Chiras

Section 21-2

The uterus and fallopian tubes are hollow muscular organs that connect the vagina, the site of sperm deposition, with the ovaries where the ova are produced.

Notes

Notes

Fifth Edition Human Biology Daniel D. Chiras — Section 21-2

The ovaries produce ova and release them during ovulation.

- Female germ cells are housed in follicles in the ovary.
- A follicle consists of a germ cell and an investing layer of follicle cells.
- A dozen or so follicles enlarge during each menstrual cycle.
- In humans, usually only one follicle makes it to ovulation.

Fifth Edition Human Biology Daniel D. Chiras — Section 21-2

- The oocyte and an investing layer of follicle cells are released during ovulation and drawn into the uterine tubes.
- The follicle then collapses and is converted into a corpus luteum (CL), which releases estrogen and progesterone.
- In the absence of fertilization, the CL degenerates.

Fifth Edition Human Biology Daniel D. Chiras — Section 21-2

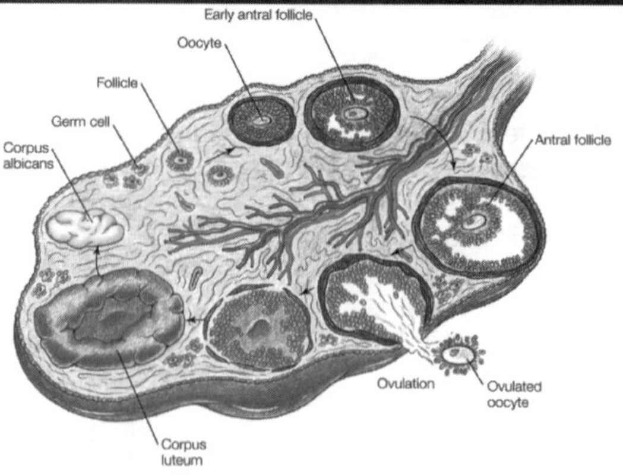

Figure 21- 9 Structure of the Ovary

Fifth Edition Human Biology Daniel D. Chiras — Section 21-2

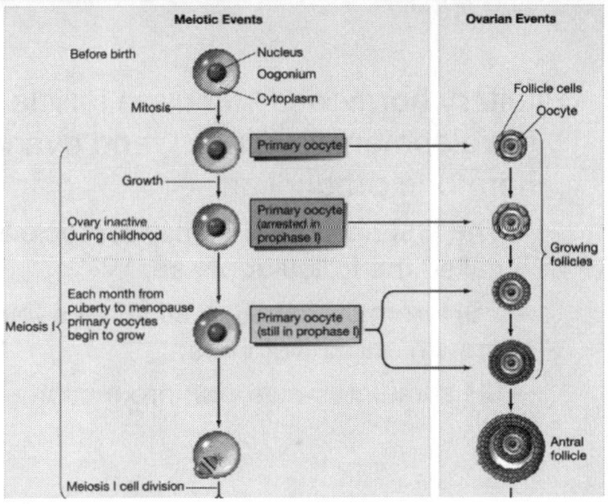

Figure 21-10 Meiosis I

Fifth Edition Human Biology Daniel D. Chiras — Section 21-2

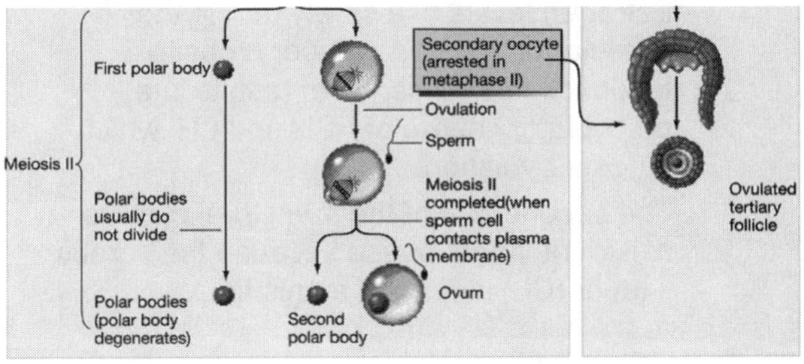

Figure 21-10 Meiosis II

Fifth Edition Human Biology Daniel D. Chiras — Section 21-2

The menstrual cycle consists of cyclic changes in hormone level, the uterus, and the ovaries.

- Changes in the hormones from the pituitary gland cause changes in the ovary. Ovarian hormones cause changes in the uterus.
- The hormonal, ovarian, and uterine changes constitute the menstrual cycle.

Notes

Notes

Pituitary hormones stimulate follicle development, ovulation, and ovarian hormone production.

- The first half of the menstrual cycle is called the follicular phase.
- FSH from the pituitary stimulates follicle growth and development.
- LH stimulates estrogen production.

Fifth Edition Human Biology Daniel D. Chiras — Section 21-2

- Estrogen levels rise slowly during the follicular phase, then trigger a positive feedback mechanism that results in a preovulatory surge of FSH and LH, which triggers ovulation.
- The second half of the menstrual cycle is called the luteal phase because the corpus luteum (CL) is formed at this time.

Fifth Edition Human Biology Daniel D. Chiras — Section 21-2

The uterine lining, or endometrium, also undergoes cyclic changes during the menstrual cycle.

- During the menstrual cycle, ovarian hormones stimulate growth of the uterine lining, which is necessary for successful implantation.
- If fertilization does not occur, the uterine lining is sloughed off during menstruation.
- Menstruation is triggered by a decline in ovarian estrogen and progesterone.

Fifth Edition Human Biology Daniel D. Chiras — Section 21-2

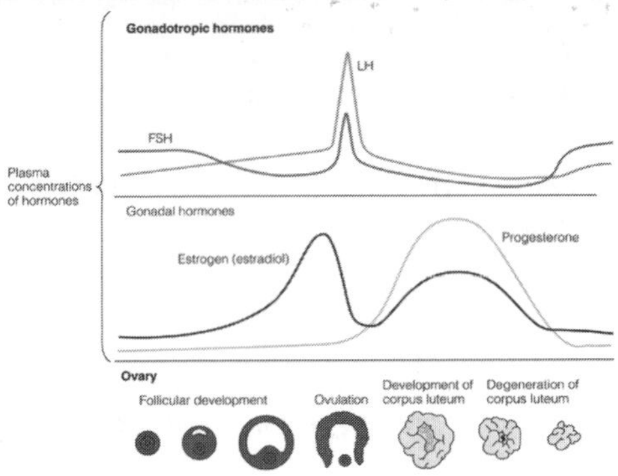

Figure 21-11a, b The Menstrual Cycle

Fifth Edition Human Biology Daniel D. Chiras — Section 21-2

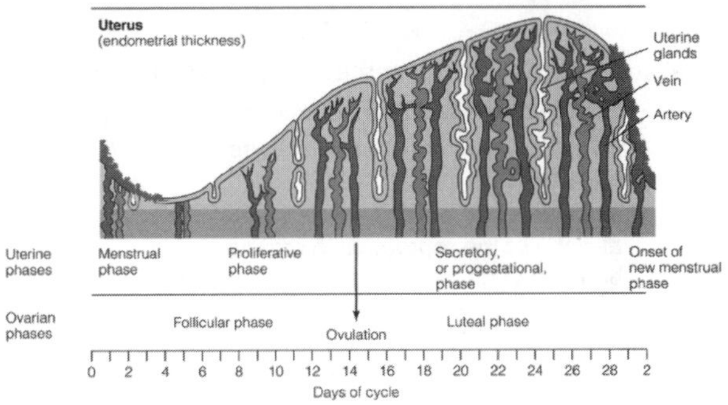

Figure 21-11c The Menstrual Cycle

Fifth Edition Human Biology Daniel D. Chiras — Section 21-2

Fertilization prevents the uterine lining from being sloughed off.

- If fertilization occurs, human chorionic gonadotropin (HCG) from the embryo maintains the CL for approximately 6 months.

Notes

Notes

Human Biology, Fifth Edition — Daniel D. Chiras — Section 21-2

The female sex steroids, estrogen and progesterone have numerous effect.

Premenstrual syndrome (PMS) is a condition afflicting many women.

Estrogen secretion at puberty stimulates:

- Follicle development in the ovaries, external growth of genitalia and breasts, growth of uterus, uterine tubes and vagina, rapid bone growth, and fat deposition.

Human Biology, Fifth Edition — Daniel D. Chiras — Section 21-2

Menopause is the cessation of menstruation.

- Estrogen levels slowly decline as a woman ages.
- Ovulation and menstruation cease between the ages of 45 and 55.
- The decline in estrogen levels results in the atrophy of the reproductive organs and in behavioral disturbances.

Human Biology, Fifth Edition — Daniel D. Chiras

Section 21-3 Birth Control

Contraceptive measures help prevent pregnancy.

- Abstinence
- Sterilization
- The pill
- Intrauterine device
- The diaphragm, condom, and sponge
- Withdrawal
- Spermicidal chemicals
- The rhythm method

Abortion is the surgical termination of pregnancy.

Fifth Edition Human Biology Daniel D. Chiras **Section 21-3**

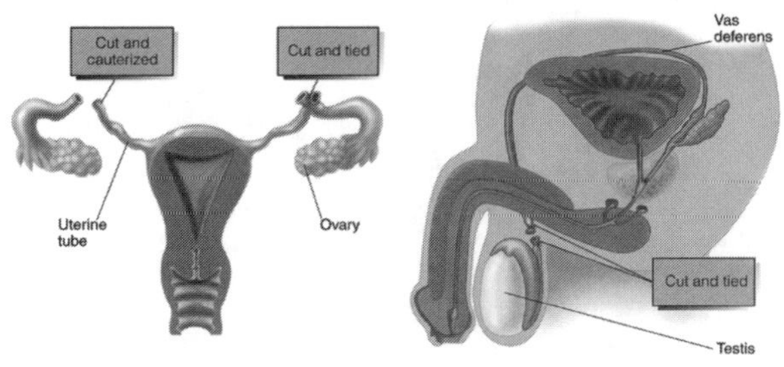

Figure 21-13
Sterilization Methods

Fifth Edition Human Biology Daniel D. Chiras

Section 21-4 Infertility

Infertility is fairly common among reproductive age couples.

The inability to conceive may be due to a variety of reasons.

Fifth Edition Human Biology Daniel D. Chiras

Section 21-5 Health and Homeostasis

A variety of chemicals in the environment may be causing low sperm count of American men.

Notes

Notes

Fifth Edition Human Biology Daniel D. Chiras

Sexually Transmitted Diseases

Sexually transmitted diseases (STDs) are caused by bacteria and viruses transmitted by sexual contact.

- Symptoms of some STDs affect the whole body, not just the reproductive tract.
- Some STDs do not have obvious symptoms. This makes transmitting the disease unintentionally more likely.

Fifth Edition Human Biology Daniel D. Chiras

Gonorrhea

Gonorrhea is a bacterial infection that can spread to many organs.

- Gonorrhea often causes no symptoms.
- When symptoms do occur, they appear 1-14 days after sexual contact.
- Gonorrhea is treated with antibiotics. If untreated, the disease can spread to other organs.

Notes

Syphilis

Syphilis is caused by a bacterium and can be extremely debilitating if untreated.

- Syphilis can be transmitted through sexual contact and from mother to child through the placenta.
- If untreated, syphilis goes through three stages.

Stage 1

- A small red sore develops, usually in the genital area.
- About 6 weeks after the sore heals, the person has a fever, headache, and loss of appetite.
- A rash appears on the body and lymph nodes are swollen.

Stage 2

- Symptoms disappear, sometimes for years.

Stage 3

- Patients lose both their sense of balance and sensation in their legs.
- Paralysis, senility, and insanity occur as the disease progresses.
- Some people go blind.
- Ulcers may form on the skin or internal organs.
- Syphilis can be treated with antibiotics only if the disease is detected early.

Fifth Edition Human Biology Daniel D. Chiras

Chlamydia

Chlamydial infections are extremely common among college students.

- Chlamydia is bacterial infection that can cause infertility if untreated.
- Symptoms include painful urination and a discharge from either the penis or vagina.
- There may be no symptoms of chlamydia.
- Antibiotics are an effective treatment for this disease.

Fifth Edition Human Biology Daniel D. Chiras

Genital Herpes

Genital herpes is caused by a virus, is extremely common, and essentially incurable.

- Early symptoms include pain, tenderness, and itchiness of the external genitalia. Later, painful blisters appear.
- Herpes can be transmitted during sexual contact only when the blisters are forming or present.
- Herpes cannot be cured, but antiviral drugs can help suppress outbreaks.

Fifth Edition Human Biology Daniel D. Chiras

Nongonoccal Urethritis

NGU is an extremely common disease caused by several types of bacteria.

- NGU is the most common type of STD.
- Although generally less dangerous than other STDs, NGU can cause sterility.
- Antibiotics are an effective treatment, but patients should seek treatment quickly to avoid serious complications.

Notes

Notes

Fifth Edition Human Biology Daniel D. Chiras

Genital Warts

Genital warts are caused by human Papillomavirus (HPV).

- The majority of Americans carry a virus called papillomavirus.
- When the immune system is weak, genital warts may occur after transmission via sexual contact.
- Chemical or surgical removal of warts is possible, but recurrences are common.

Fifth Edition Human Biology Daniel D. Chiras

Preventing STDs

Abstinence is the best protect against STDs.

Condoms are also effective protection against transmission of bacteria and viruses.

Notes

Chapter 22

Human Development and Aging

Section 22-1 Fertilization

In humans, fertilization usually occurs in the upper third of the oviduct.

Very few sperm make it to the site of fertilization.

Sperm travel by swimming and muscular contractions in the walls of the uterus and uterine tube.

Sperm contain enzymes that help them penetrate the protective layers surrounding the egg.

Only one sperm can fertilize an egg.
- Further sperm penetration is blocked.
- The sperm is engulfed by the oocyte.

Notes

Fifth Edition Human Biology Daniel D. Chiras — Section 22-1

Figure 22-2a Fertilization

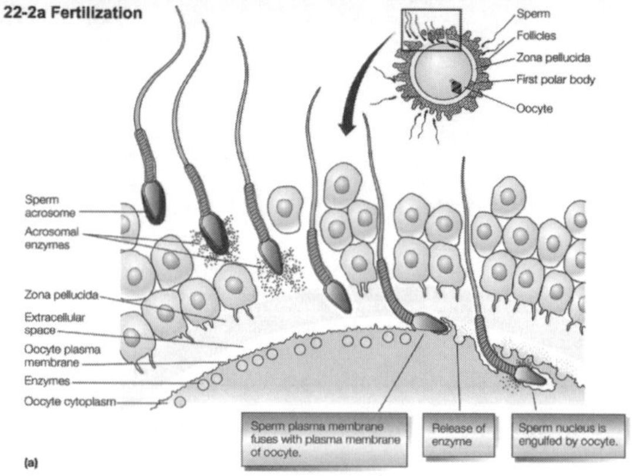

Fifth Edition Human Biology Daniel D. Chiras — Section 22-1

Sperm and egg nuclei join to create a zygote with 46 chromosomes

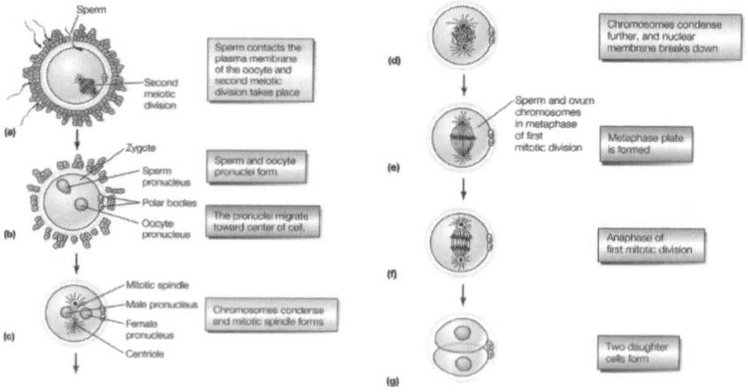

Figure 22-3

Fifth Edition Human Biology Daniel D. Chiras

Section 22-2 Pre-embryonic Development

Pre-embryonic development begins at fertilization and ends just after implantation.

- The zygote undergoes rapid cellular division and forms a morula.
- The morula arrives in the uterus 3–4 days after fertilization and is converted into a blastocyst, which implants in the uterine wall 2–3 days later.

Notes

Fifth Edition
Human Biology Daniel D. Chiras
Section 22-2

- The blastocyst consists of a clump of cells.
  - The inner cell mass (ICM), becomes the embryo.
  - The trophoblast, gives rise to the embryonic portion of the placenta.
  - While the placenta is forming, a layer of cells from the ICM of the blastocyst separates from it and forms the amnion.
  - The amnion fills with fluid and enlarges during embryonic and fetal development.

Fifth Edition
Human Biology Daniel D. Chiras
Section 22-2

The embryo embeds in the wall of the uterus.
- Implantation occurs 6-7 days after fertilization.
- If the endometrium is not ready, or unhealthy, implantation does not occur.

Implantation occasionally occurs outside of the uterus.
- An ectopic pregnancy is one in which the embryo implants in the uterine tubes or outside the reproductive tract.

Fifth Edition
Human Biology Daniel D. Chiras
Section 22-2

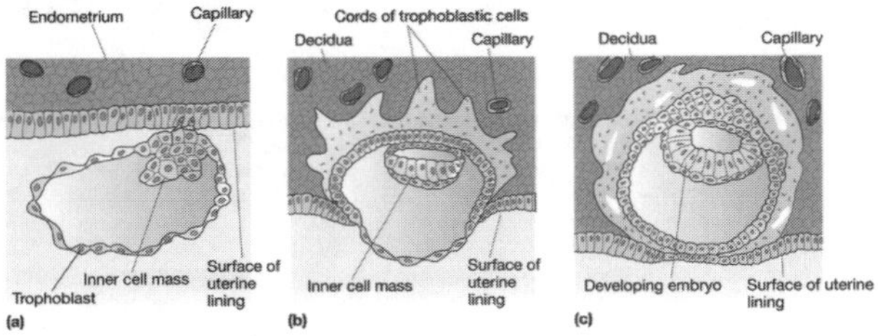

Figure 22-5 Implantation

Notes

Fifth Edition Human Biology Daniel D. Chiras — Section 22-2

Placental hormones are essential to reproduction. Three placental hormones include:

- Human chorionic gonadotropin (HCG)
- Estrogen
- Progesterone

The amnion forms from the inner cell mass (ICM).

- The amniotic cavity forms between the ICM and the amnion.

Fifth Edition Human Biology Daniel D. Chiras

Section 22-3 Embryonic Development

Embryonic development ends when the organs are more or less formed.

- After the amnion forms, the cells of the ICM differentiate into the three germ cell layers: ectoderm, mesoderm, and endoderm.
- The formation of the three primary germ layers marks the beginning of embryonic development.
- The organs develop from the three basic tissues during organogenesis.

Fifth Edition Human Biology Daniel D. Chiras — Section 22-2

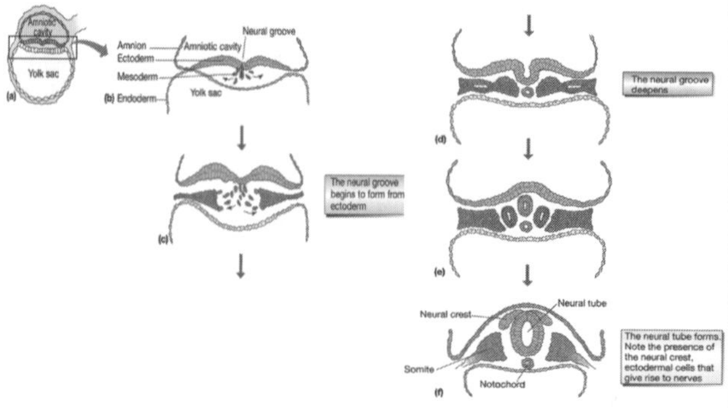

Figure 22-8 Formation of the Spinal Cord

Fifth Edition Human Biology Daniel D. Chiras

Section 22-4 Fetal Development

During fetal development organs continue to develop and grow.

- Fetal development begins in the eighth week of pregnancy.

Fifth Edition Human Biology Daniel D. Chiras — Section 22-4

The fetal circulatory system contains three bypasses around the lungs and liver.

- The umbilical cord carries blood and nutrients to and from the placenta.
- The ductus venosus shunts blood from the umbilical vein directly to the inferior vena cava, bypassing the liver.
- The foramen ovale shunts blood directly from the right ventricle to the left ventricle, bypassing the lungs.
- The ductus arteriosus also shunts blood away from the lungs.

Fifth Edition Human Biology Daniel D. Chiras — Section 22-4

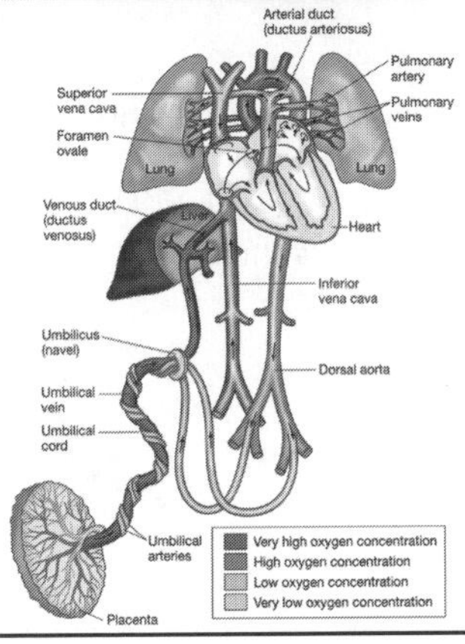

Figure 22-11a Fetal Circulation – Before Birth

Notes

Notes

Fifth Edition Human Biology Daniel D. Chiras

Section 22-5 Birth Defects and Spontaneous Abortion

Birth defects include structural and functional defects that appear in the newborn.

- About 31 percent of all successful fertilizations end in a miscarriage.
- About 10-12 percent of all newborns have birth defects.
- Birth defects arise from chemical, biological, and physical agents known as teratogens.
- The effect of teratogenic agents is related to the time of exposure, the nature of the agent, and the dose.

Fifth Edition Human Biology Daniel D. Chiras

Section 22-6 Childbirth and Lactation

Uterine muscle contractions are stimulated by a change in hormonal levels.

- It's believed that fetal oxytocin stimulates the release of prostaglandins by the placenta.
- Oxytocin and prostaglandins stimulate contractions in the sensitized uterine musculature.

Fifth Edition Human Biology Daniel D. Chiras — Section 22-6

Childbirth occurs in three stages.

- The dilation
- The expulsion
- The placental

Fifth Edition Human Biology Daniel D. Chiras — Section 22-6

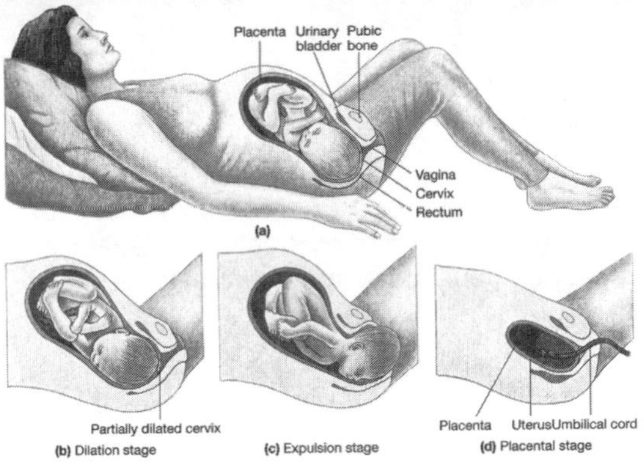

Figure 22-15 Stages of Labor

Fifth Edition Human Biology Daniel D. Chiras — Section 22-6

Milk production, or lactation, is controlled by the hormone prolactin.

- Milk production does not begin until 2–3 days after birth.
- Before milk production begins, the breasts produce small quantities of a protein-rich fluid, colostrum.
- A newborn can subsist on colostrum for the first few days
- The baby derives antibodies from colostrum that help protect it from bacteria.
- Suckling causes a surge in prolactin secretion.
- Each surge stimulates milk production needed for the next feeding.

Fifth Edition Human Biology Daniel D. Chiras

Section 22-7 Aging and Death

Aging is the progressive deterioration of the body's structure and function.

Notes

Notes

Aging results as a result of a decline in cell numbers and their function.

Aging is often associated with certain diseases, but the likelihood of contracting a disease of old age depends in large part on life-style.

Fifth Edition Human Biology Daniel D. Chiras Section 22-7

Death may be defined by the loss of brain or heart and pulmonary function.

Death can also result from aging, traumatic injury, or infectious disease.

- The moment of death is difficult to define.

Fifth Edition Human Biology Daniel D. Chiras

Section 22-8 Health and Homeostasis

Homeostasis is essential for normal reproduction.

Radiation from video display terminals might help cause miscarriages, but further research is needed to verify the connection.

Notes

Section 23-1 The Evolution of Life: An Overview

The evolution of life can be divided into three phases:
- Chemical evolution
- Cellular evolution
- The evolution of multicellular organisms

The formation of organic molecules from inorganic molecules is chemical evolution.
- Chemical evolution possibly began 4 billion years ago.
- Inorganic molecules (from the atmosphere combined to form small organic molecules.
- Energy needed to drive the reaction may have come from sunlight or other sources.
- The small organic molecules combined to form polymers.

Notes

Fifth Edition Human Biology Daniel D. Chiras — Section 23-1

The theory of chemical evolution is supported by several convincing experiments.

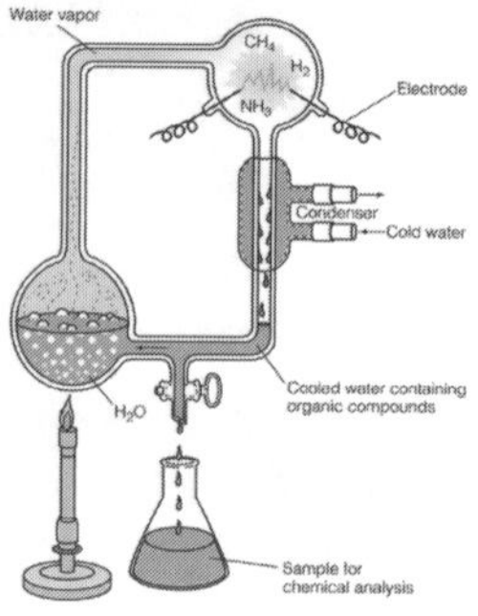

Figure 23-1 Miller's Apparatus

Fifth Edition Human Biology Daniel D. Chiras — Section 23-1

The earliest cells were probably aggregations of polymers that acquired nutrients directly from the environment.

- These aggregates, called protocells, may have synthesized some of their own molecules.
- The very first cells probably acquired nutrients from the environment.
- The earliest autotrophs were probably bacteria that acquired energy from chemicals in the environment.
- Cellular evolution is the evolutionary development of prokaryotic cells from cell precursors.

Fifth Edition Human Biology Daniel D. Chiras — Section 23-1

Eukaryotes probably arose by endosymbiotic evolution.

- Eukaryotes evolved about 1.2 billion years ago.
- The first eukaryotes were single-celled organisms with nuclei and distinct organelles.

Eukaryotes gave rise to multicellular organisms.

Fifth Edition Human Biology Daniel D. Chiras — Section 23-1

The evolution of multicellular organisms occurred rather rapidly compared with previous stages.

- The evolution of eukaryotes opened the door for the evolution of multicellular organisms.
- The evolution of multicellular life began in the sea.
- Plants and animals that inhabited the land arose from the seas.
- The first animals were probably arthropods, scorpion-like creatures.

Fifth Edition Human Biology Daniel D. Chiras

Section 23-2 How Evolution Works

Genetic variation is the raw material of evolution.

Changes in species occur because of changes in genetic material.

- Genetic variation in a species arises from several processes including:
  - Mutations
  - new combinations resulting from sexual reproduction.

Fifth Edition Human Biology Daniel D. Chiras — Section 23-2

Natural selection is a process by which organisms become better adapted to their environment.

- Natural selection is a process in which slight variations, if useful, are preserved.
- Abiotic and biotic factors contribute to natural selection.

Natural selection ensures that the fittest organisms survive and reproduce.

- Fitness is a measure of reproductive success.

Notes

Notes

Fifth Edition
Human Biology Daniel D. Chiras

Section 23-3 The Evidence Supporting Evolution

The fossil record yields some of the best supporting evidence for the existence of evolution.

Common anatomical features in different species support the theory of evolution.

- Comparing homologous structures permits scientists to determine evolutionary relationships.

Fifth Edition
Human Biology Daniel D. Chiras

Section 23-3

The common biological makeup of organisms also supports the theory of evolution.

Studies of common developmental patterns supports the theory of evolution.

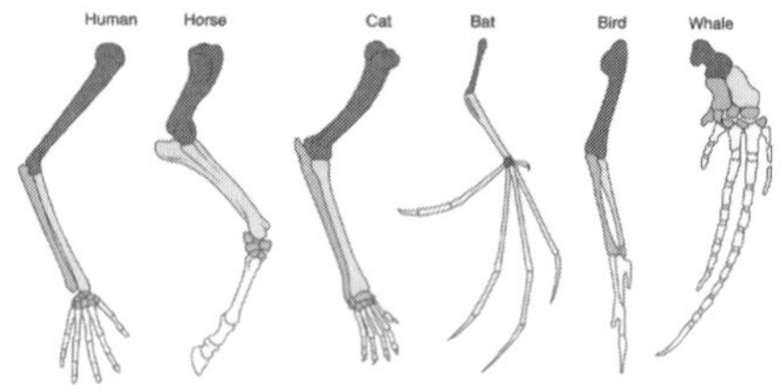

Figure 23-5

Fifth Edition
Human Biology Daniel D. Chiras

Section 23-3

Experimental evidence supports the theory of evolution by natural selection.

The study of biogeography helps explain evolution.

- Biogeography is the study of the distribution of plants and animals worldwide.

Fifth Edition Human Biology Daniel D. Chiras

Section 23-4 Early Primate Evolution

Humans belong to the order called primates, which includes two suborders:

- Prosimians (premonkeys
- Anthropoids (monkeys, apes, and humans.

The prosimians probably evolved from tree-dwelling mammals that lived during the age of the dinosaurs.

Fifth Edition Human Biology Daniel D. Chiras — Section 23-4

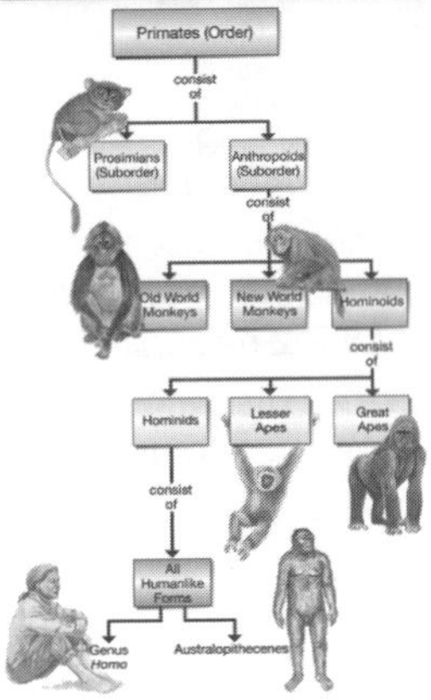

Figure 23-8 The Evolution of Primates

Fifth Edition Human Biology Daniel D. Chiras — Section 23-4

Primates are characterized by grasping hands, forward-looking eyes, and large brains.

The fossil evidence of primate evolution is limited, making it difficult to determine the exact progression of early human ancestors.

Notes

Notes

Fifth Edition Human Biology Daniel D. Chiras

Section 23-5 Evolution of the Australopithecines

The earliest hominids were members of the genus *Australopithecus*.

- The oldest known australopithecine skeleton, named Lucy, is believed to be about 3.5 million years old.
- The skeleton belongs to a group called *Australopithecus afarensis* and stood about 3 feet high.

Fifth Edition Human Biology Daniel D. Chiras — Section 23-5

About 3 million years ago, *A. afarensis* was replaced by *A. africanus*.

A. robustus appeared approximately 2.3 million years ago, and *A. boisei* appeared approximately 2.2 millions ago.

Australopithecines disappeared about 1 million years ago; no one is sure why.

Fifth Edition Human Biology Daniel D. Chiras — Section 23-5

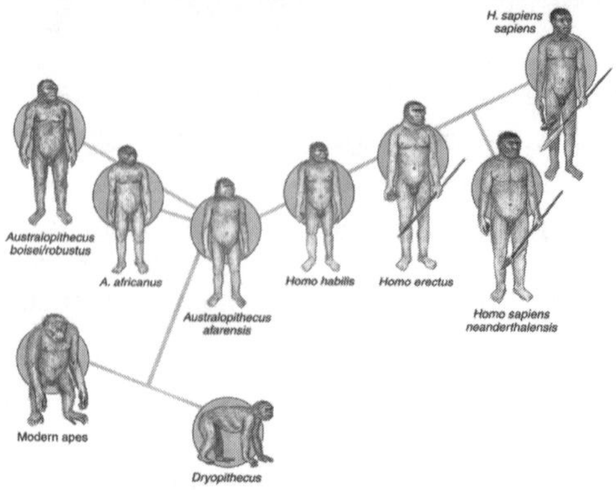

Figure 23-10 Evolution of *Homo sapiens*

Fifth Edition
Human Biology Daniel D. Chiras

Section 23-6 Evolution of the Genus *Homo*

The first truly humanlike creatures were *Homo habilis*.

Homo habilis gave rise to *Homo erectus*.

- *H. erectus* moved from Africa to Europe and Asia.
- *Homo erectus* is believed to be the direct ancestor of *Homo sapiens*.

Fifth Edition
Human Biology Daniel D. Chiras — Section 23-6

Modern humans belong to *Homo sapiens*.

- *Homo sapiens* emerged about 300,000 years ago
- *Homo sapiens* consists of two subspecies:
  - *Homo sapiens neanderthalensis* (the Neanderthals
  - *Homo sapiens sapiens*
- Cro-Magnons are the earliest known members of *H. sapiens sapiens*.
- The Neanderthals disappeared about 40,000 years ago.

Fifth Edition
Human Biology Daniel D. Chiras — Section 23-6

Human races result from variations caused by geographic separation.

Figure 23-17

Notes

Notes

Fifth Edition Human Biology Daniel D. Chiras

Section 23-7 Health and Homeostasis

Homeostasis can be upset by pesticides.

- Humans have devised ways to use and modify the environment to meet their needs.
- Some human modifications, like pesticides, have damaged the environment's homeostatic mechanisms.
- Weeds and insects can become resistant to herbicides and pesticides, requiring stronger chemicals to kill them.

Notes

Chapter 24

Ecology and the Environment

Fifth Edition Human Biology Daniel D. Chiras

Section 24-1 An Introduction to Ecosystems

Ecology is the study of the interactions between organisms, including humans, and their environment.

Courtesy of the NPS

Fifth Edition Human Biology Daniel D. Chiras — Section 24-1

The biosphere is the zone in which all life exists on Earth.

- The biosphere extends from the bottom of the oceans to the top of the highest mountains.
- The biosphere is a closed system in which materials are recycled over and over.
- The only outside contribution is sunlight, which powers virtually all biological processes.

Notes

Fifth Edition Human Biology Daniel D. Chiras — Section 24-1

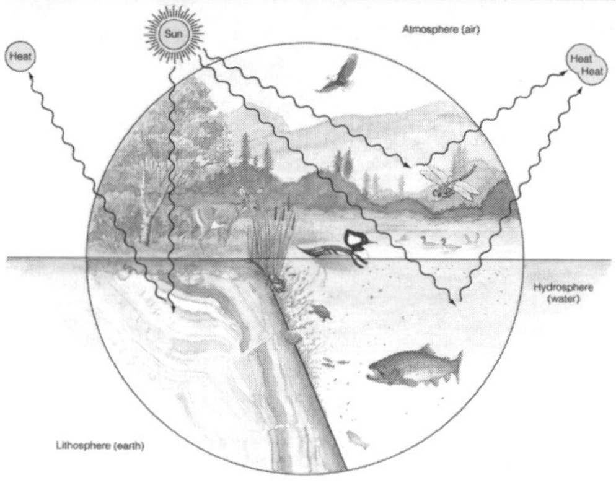

Figure 24-1
The Biosphere

Fifth Edition Human Biology Daniel D. Chiras — Section 24-1

The biosphere is divided into biomes and aquatic life zones.
- Each biome and aquatic life zone has characteristic plant and animal life.
- The biomes have characteristic climates.

Fifth Edition Human Biology Daniel D. Chiras — Section 24-1

Ecosystems consist of organisms and their environment.
- Ecosystems consist of two basic components: abiotic and biotic.
  - Abiotic components are the physical and chemical factors needed for life.
  - Biotic components are the organisms that live in an ecological system.
- A group of organisms of the same species living in a specific region constitutes a population.
- Two or more populations occupying that region form a community.

Fifth Edition Human Biology Daniel D. Chiras — Section 24-1

Habitat is the physical space where an organism lives.

An organism's niche consists of all of an organism's relationships in an ecosystem.

Fifth Edition Human Biology Daniel D. Chiras

Section 24-2 Ecosystem Function

Producers generate nutrients consumed by all other organisms.

- Producers are organisms that synthesize organic materials from sunlight, carbon dioxide, and water.
- The major producers are the plants, photosynthetic protists, and photosynthetic bacteria.
- Consumers are organisms dependent on producers and other organisms for food.
    - Four types of consumers are present: herbivores, carnivores, omnivores, and detritivores.

Fifth Edition Human Biology Daniel D. Chiras — Section 24-2

Organisms are part of food webs.

- A food chain represents a feeding relationship in an ecosystem.
- There are two general types of food chains: grazer and decomposer.
- Energy and nutrients both flow through food webs but in very different ways.

Notes

Notes

Fifth Edition Human Biology Daniel D. Chiras — Section 24-2

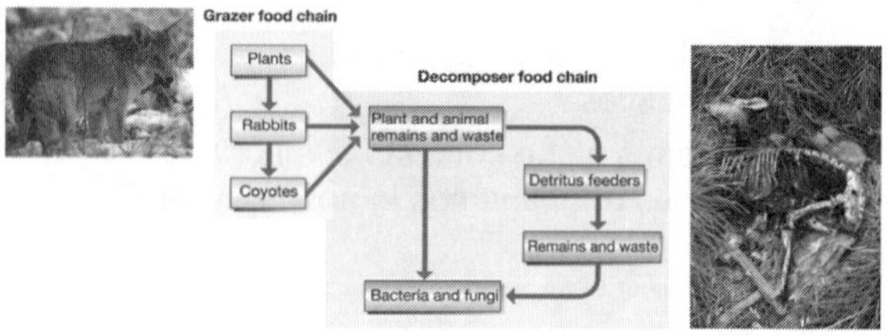

Figure 24-4 Food Chains

Fifth Edition Human Biology Daniel D. Chiras — Section 24-2

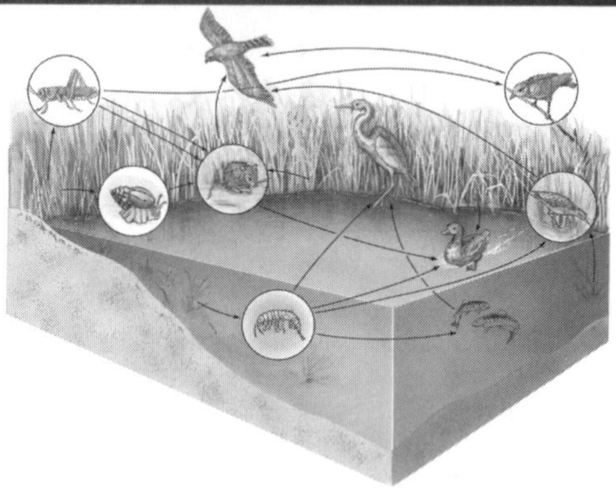

Figure 24-5 Food Web

Fifth Edition Human Biology Daniel D. Chiras — Section 24-2

The organisms of a food chain exist on different trophic levels.

- In a grazer food chain, plants (producers are on the first trophic level; herbivores are on the second; carnivores are on the third.
- An energy pyramid is a plot of the chemical energy of the trophic levels.

Fifth Edition
Human Biology Daniel D. Chiras — Section 24-2

Nutrient cycles consist of two phases, the organismic and the environmental.

- In the organismic phase, nutrients are found in the biota.
- In the environmental phase, a nutrient exists in the air, water or soil, or in two or more.
- Three important nutrient cycles are the
  - Water cycle
  - carbon cycle
  - nitrogen cycle

Fifth Edition
Human Biology Daniel D. Chiras — Section 24-2

The water cycle, or hydrological cycle, collects, purifies, and distributes water throughout the planet.

- Evaporation and precipitation run the water cycle.

Fifth Edition
Human Biology Daniel D. Chiras — Section 24-2

The carbon cycle recycles carbon throughout the planet.

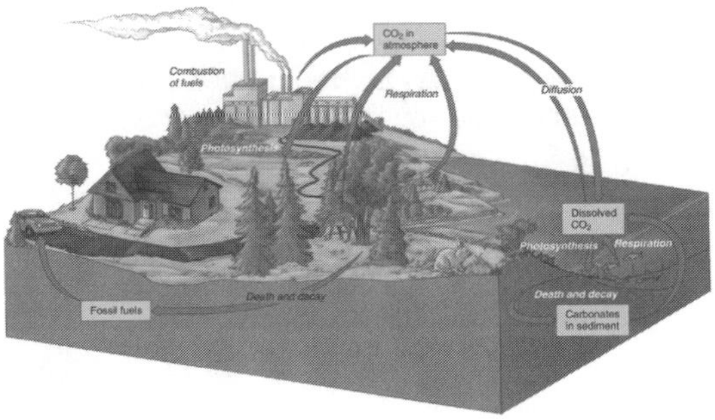

Figure 24-8

Notes

Notes

Fifth Edition Human Biology Daniel D. Chiras — Section 24-2

The nitrogen cycle recycles nitrogen throughout the planet.

- Nitrogen fixation is the conversion of atmospheric nitrogen to ammonia.

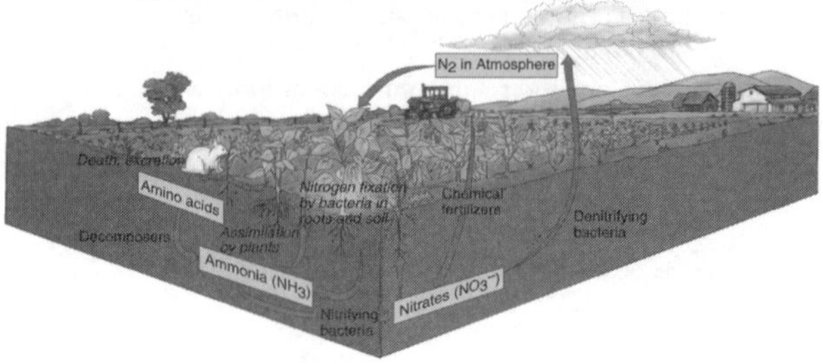

Figure 24-9

Fifth Edition Human Biology Daniel D. Chiras — Section 24-2

Succession is the progressive development of biological communities.

- Primary succession occurs where no community previously existed.
- Secondary succession occurs where a community was destroyed by natural or human events.

Fifth Edition Human Biology Daniel D. Chiras

Section 24-3 Overshooting the Earth's Carrying Capacity

Carrying capacity is the number of organisms an ecosystem can sustain indefinitely.

It is determined by

- food production.
- resource supplies.
- the capacity of the environment to assimilate or destroy waste products of organisms.

Fifth Edition Human Biology Daniel D. Chiras — Section 24-3

In many places, the human population is exceeding food production.

Many resources are in short supply and will be depleted in the near future.
- Nonrenewable resources are finite.
- Renewable resources can replenish themselves naturally.

Pollution from human activities exceeds the environment's assimilative capacity.

Fifth Edition Human Biology Daniel D. Chiras

Section 24-4 Overpopulation
Problems and Solutions

Overpopulation is a condition in which populations exceed the ability of the environment to supply resources and/or assimilate wastes.

Overpopulation is a problem in virtually all countries, rich and poor.
- The human population is growing exponentially.

Reducing population growth will help reduce environmental problems.

Fifth Edition Human Biology Daniel D. Chiras

Section 24-5 Resource Depletion
Eroding the Prospects of All Organisms

Humanity is destroying the world's forests, but massive replanting efforts could reduce, even halt, the problem.

Notes

Notes

Fifth Edition Human Biology Daniel D. Chiras — Section 24-5

Soil erosion, like deforestation, is also a worldwide phenomenon.

- The destruction of productive soils threatens the long-term prospects for food production.
- Soil conservation and population control measures can help ensure an adequate supply of soil

Fifth Edition Human Biology Daniel D. Chiras — Section 24-5

Many areas of the world are facing water shortages or will soon face them as the human population and demand for water increase.

Most students alive today will see the end of oil in their lifetimes.

Fifth Edition Human Biology Daniel D. Chiras

Section 24-6 Pollution

Our waste is overwhelming nutrient cycles, poisoning other species (and ourselves.

Global warming results from the release of carbon dioxide and other air pollutants.

- Carbon dioxide is a greenhouse gas affecting global warming.

The environmental impacts of global warming could be severe.

Solving global warming requires massive action, and soon.

Fifth Edition Human Biology Daniel D. Chiras — Section 24-6

Large portions of the world are threatened by acid rain and snow.

- Power plants, factories, automobiles, and other sources release sulfur dioxide and nitrogen dioxide as gaseous pollutants.
- In the atmosphere, they are converted to sulfuric and nitric acid, respectively.
- Acids fall from the sky in wet and dry deposition.

Fifth Edition Human Biology Daniel D. Chiras — Section 24-6

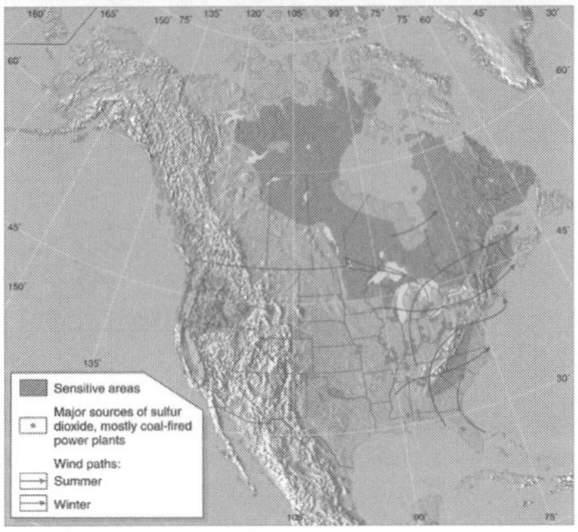

Figure 24-20 Acid Deposition and Sensitive Areas

Fifth Edition Human Biology Daniel D. Chiras — Section 24-6

The ozone layer is endangered by human pollutants.

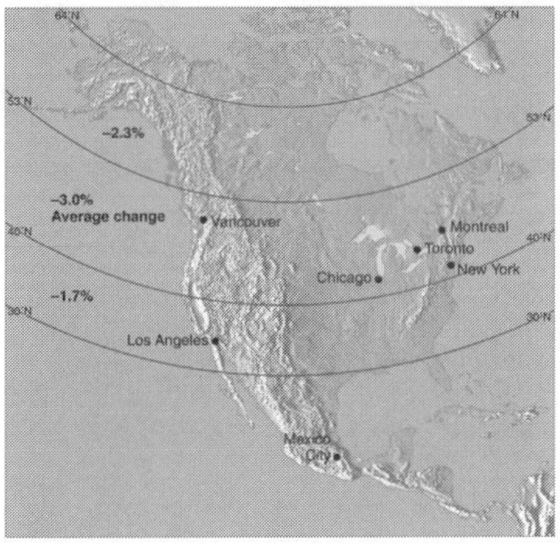

Figure 24-22

Notes

Notes

Fifth Edition Human Biology Daniel D. Chiras

Section 24-7 Health and Homeostasis

A sustainable society is one that lives within the carrying capacity of the environment.

A sustainable society is built on five principles:

- Conservation
- Recycling
- Renewable resources
- Restoration
- Population control